Sheep production: Science into practice

Longman Handbooks in Agriculture

Whittemore: *Lactation of the Dairy Cow*
Whittemore: *Pig Production: the scientific and practical principles*
Speedy: *Sheep Production: Science into practice*
Hunter: *Reproduction of Farm Animals*
Leaver: *Milk Production*

Sheep production

Science into practice

Andrew W. Speedy

Lecturer in Animal Science, Department of Agricultural and Forest Science, University of Oxford

 Longman Scientific & Technical

Longman Scientific & Technical
Longman Group UK Limited,
Longman House, Burnt Mill, Harlow
Essex CM20 2JE, England
and Associated Companies throughout the world

Published in the United States of America
by Longman Inc., New York

First published 1980
Reprinted 1982, 1985, 1989

British Library Cataloguing in Publication Data

Speedy, Andrew
  Sheep production. – (Longman handbooks in
  agriculture).
  1. Sheep
  I. Title
  636.3'08   SF375   79-41352

ISBN 0-582-45582-0

Produced by Longman Singapore Publishers (Pte) Ltd.
Printed in Singapore

Contents

Preface

In the last 30 years, science has rearmed the sheep farmer with a multitude of new developments and materials. In the fields of pasture management, nutrition, breeding and health control, these represent major advancements in the potential for sheepmeat production.

To realize the full benefits of scientific research and development, new techniques must be incorporated into practical farming systems and, within the mixed farm, integrated with the demands of other enterprises in the whole farming system. This book describes the technical details and the practical application of sheep production systems within the financial constraints and scale of modern farming. It provides students with up-to-date knowledge on management, breeding and feeding of sheep and helps practical farmers to improve the productivity and profitability of their flocks.

Much of the technical information reported comes from the progressive work of scientists at the Rowett Research Institute, Hill Farming Research Organisation, Animal Breeding Research Organisation and other centres throughout the UK and elsewhere. Integration of this fundamental work into practical systems has been the special role of the Scottish Agricultural Colleges, the Agricultural Development and Advisory Service and the Meat and Livestock Commission (MLC). Contact with many colleagues in these organizations has provided the basis for this book. I am grateful to all those people who have provided the material in published form or in discussion and hope that the result does justice to the high standard of their contributions to the field. *Farmers' Weekly* kindly gave me permission to reproduce the following Figures: 5.7, 6.3, 6.4, 6.5, 6.7, 6.8, 12.3, 12.9, 13.2, 13.3, 13.4.

I wish to record particular thanks to the MLC for support in my own work over the last 10 years and all my friends and colleagues at Edinburgh.

Finally, I would thank my wife for her support throughout.

Andrew Speedy
Edinburgh 1979

The role of sheep production in agriculture

1

The major advantage of sheep in agricultural systems is their ability to utilize pasture to produce saleable meat and wool. In hill and upland areas, they use land that would otherwise be of little value for agricultural purposes and, in the lowlands, they utilize the grass break in an arable rotation. The principal role of sheep is therefore as a grazing animal, and the aim of efficient sheep production is to maximize output from pasture, with some help from conserved fodder and forage crops.

In the UK there are seven million hectares of hill and upland land, the majority of which is unsuitable for cropping except for limited production of livestock feed where cultivation is possible. However, this land supports over seven million breeding ewes and produces about 120 000 tonnes of sheepmeat and 18 000 tonnes of wool annually. The performance of sheep under extensive production systems is modest compared to more intensive livestock systems, but it is production achieved at relatively low cost in terms of land, energy and human resources. If the output from hill and upland sheep were to be increased, it would have a very significant impact on national agricultural production.

Although sheep are not as profitable as crop production in fertile lowland areas, they contribute to the restoration of soil fertility while at the same time producing saleable products from grass within the arable rotation. The importance of a grass break in intensive cereal production has become increasingly recognized. In Britain, it is estimated that about a million acres of cereals are suffering declining yields as a result of persistent single-cropping. The most satisfactory solution is a grass break utilized by ruminant livestock. The introduction of beef or dairy cattle involves high capital investment, whereas the capital required for sheep is relatively low.

Sheep production from lowland grass has considerable potential for improvement. Current use of fertilizer on grass for sheep is low – Meat and Livestock Commission (MLC) records suggest that over 70 per cent of farmers in the UK use less than 100 kg nitrogen (N)/ha. Yet it is well established that grass production increases linearly up to about 350 kg N/ha and this level is currently exploited by many dairy farmers. Hence the opportunity exists to increase sheep production from the relatively small area of lowland grass.

In the lowlands the performance of sheep can be considerably higher than in the hills and the production of two lambs per ewe, finished off grass, is well within the scope of current technology.

It is also possible to combine hill and lowland resources so that poor land is used for the maintenance of the breeding flock and supplies lambs to finish on lowland farms. Forage crops can produce very high yields of dry matter per hectare and can be used to finish large numbers of lambs when this is not possible on hill farms. Finishing lambs on a root break also restores fertility to arable land and utilizes the potential of both the hills and lowlands for meat production.

In arid areas of the world, a similar combination of extensive range conditions for the breeding flock and intensive finishing of lambs in feedlots enables greater output to be achieved than is possible when lambs are produced on the range alone.

Targets for sheep production

Within the various environments in which sheep are kept there is increasing pressure to improve profits and maximize output from the limited resources. Output must increase (without resorting to the use of more expensive cereal feeds) either by increasing production per animal and/or by increasing the number of animals carried on the given land resources.

There is considerable scope for increasing the productivity of sheep in both hill and lowland systems. Surveys of current sheep performance in Scotland, for example, show a wide variety of results between farms (Table 1.1) so it is clearly possible to improve output on many farms.

Targets of 100 per cent for hill sheep and 150–200 per cent in upland and lowland flocks are readily attainable. The MLC have shown the importance of the different factors which determine the weaning percentage (Table 1.2). A modest improvement at each level could markedly improve overall results. Much of the necessary improvement can be achieved by reducing mortality through improved nutrition and disease

Table 1.1 The range of performance in recorded commercial flocks in Scotland (source: *Sheep Package Review.* (1978) Council of the Scottish Agricultural Colleges (COSAC))

	Lambs weaned per 100 ewes
Hill farms	50–127
Upland farms	128–183
Lowland farms	108–198

Table 1.2 The physical performance of recorded sheep flocks in the UK (per 100 ewes put to the ram) (from MLC (1978) *Sheep Facts*)

	Hill	Upland	Lowland
Ewes to ram	100	100	100
Ewe deaths before lambing	2	3	2
Barren ewes	10	5	6
Productive ewes	88	92	92
Later ewe deaths	2	1	2
Total lambs born	–	139	161
Lambs born alive	–	130	150
Lambs reared	93	124	140
Lambs retained for breeding	29	12	–

control. Increased growth rates of lambs would also lead to a greater weight of weaned lamb produced, and again nutrition holds the main key to success. The main areas in which immediate improvements can be made are:

Increased numbers of lambs born per ewe;
Fewer barren ewes;
Improved lamb birthweight and consequent reduction in losses at lambing time;
Increased milk production and early lamb growth;
Increased growth rate of weaned lambs;
Reduced wastage of breeding sheep and lambs by better control of disease.

Better nutrition of ewes prior to mating and during pregnancy improves conception rates, increases the likelihood of twinning and gives heavier lambs at birth. Lamb growth rates are increased by better nutrition of both lactating ewes and lambs. Improved lamb growth rates lead to earlier slaughtering of lambs and/or higher weaning weights. Better lambing management reduces mortality of lambs, and improved health control reduces losses of ewes and lambs from disease factors as well as improving growth rates by reducing the chronic effects of disease. In particular, better control of parasites would enable feed to be used more efficiently by both ewes and lambs and exploit their full production potential.

Nutrition includes the supply of energy, protein, minerals and vitamins. Increasing the amount of feed provided must be associated with a correct balance of nutrients. The correction of trace element deficiencies, especially, has been shown to have a major impact on the productivity of sheep in all environments.

Increasing output per hectare

Besides increasing the numbers and weights of lambs produced per ewe, there is also scope for increasing the number of ewes carried per hectare of land (the stocking rate).

On hill farms the stocking rate is limited by the poor quality of the pasture and the seasonal nature of pasture growth. An increase in ewe numbers on a hill farm can be achieved by introducing supplementary feeding at critical times to meet the deficits of pasture production (or indeed the complete lack of feed in winter) and by improvement of land to increase pasture production in summer. A combination of these two methods leads to a greater carrying capacity of the hill and produces the necessary increase in output to justify the investment.

In the lowlands and better upland farms there is scope for increasing the stocking rate on existing pasture and short-term leys. The average stocking rates of sheep at higher levels of nitrogen are below the potential for efficient use of the grass produced (Fig. 1.1). At 200 kg N/ha, the average stocking rate of sheep on recorded farms is about 13 ewes/ha, whereas a stocking rate of 17.5 ewes/ha should be possible on the basis of grass production. Even average recorded stocking rates are probably above the true national average because they are not a random sample. Attempts to increase the stocking rate in the past have frequently been disappointing because increasing ewe numbers resulted in poorer lamb performance and a greater proportion of store (feeder) as opposed to finished lambs produced. There was little or no economic advantage because individual animal performance was reduced and no net increase in output was achieved. However, grass production was not the limiting factor in these cases.

The main limitation was the fact that parasite infestation increased with the extra ewes and lambs present and caused a chronic reduction in lamb performance or even acute parasitism at high stocking rates.

The opportunity to increase the summer stocking rate in the uplands and lowlands is likely to come from the application of systems designed to minimize parasite infestation while increasing grass production through the use of more nitrogen fertilizer. When sheep numbers are increased extra conserved forage and the other home-grown feeds are needed to provide the additional winter feed.

The most successful system for parasite control is the Clean Grazing System, (see Fig. 1.2). This involves the rotation of crops and grass or cattle, sheep and conservation so that ewes and lambs graze pasture which has not carried lambs or young sheep in the previous year. When ewes and lambs are grazed on clean grass, the stocking rate may be increased, using more nitrogen fertilizer, without reducing individual lamb performance. Thus, a proportional increase in output is obtained for each additional ewe carried. At Edinburgh, a clean grazing system has been in operation for the last 5 years both on College and commercial farms, and a summer stocking rate of 17.5 ewes and their lambs per hectare has been carried with excellent lamb performance and over 50 per cent

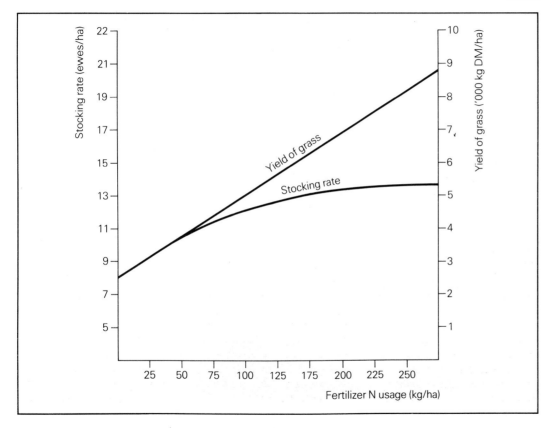

Figure 1.1 Sheep stocking rate at different levels of fertilizer N usage on recorded farms in the UK* and the yield of grass at these levels†
*MLC (1978) *Sheep Facts* Meat and Livestock Commission, Queensway House, Bletchley, Milton Keynes, MK2 2EF. †D. Reid (1972) *J. agric. Sci., Camb.* **79**, 291-307

Figure 1.2   Scottish Halfbred ewes with Suffolk cross lambs on clean grass at House O'Muir Farm, near Edinburgh

of lambs sold fat by weaning. The fertilizer used is 200 kg N/ha so that potential exists for further increasing grass production and stocking rate. Even at the level so far achieved, the stocking rate is about double that previously accepted on similar farms.

Similar improvements in stocking rate can be demonstrated for store lambs on forage crops. Crop production can be increased by earlier sowing, more productive varieties, disease control and increased fertilizer rates. With higher crop yields, a greater number of lambs can be fattened per hectare.

Intensive systems of sheep production incur a greater risk of disease problems and special attention must be paid to the control of disease. The main categories of disease are gastro-intestinal parasites, ectoparasites, pneumonias, contagious abortion, clostridial diseases, foot troubles and metabolic disorders. A routine programme of preventive medicine is essential, and vaccines, drenches, dips and footbaths are an integral part of successful flock management.

Finally, the higher costs of labour and the demands of more intensive production systems make equipment and working facilities an important consideration in modern sheep farming. This includes the design of handling pens, shearing facilities, veterinary equipment and fencing.

Every system of sheep production requires forward planning to ensure the success of the enterprise and the sheep must be integrated with the other enterprises on the farm to produce a profitable and workable whole farm business.

Grass for sheep

2

As with any grazing system, the best use of resources is achieved by fitting the pattern of grass supply to the needs of the sheep flock. In northern Europe the highly seasonal pattern of grass growth fits reasonably well with the demands of a spring lambing flock when the greatest requirement comes from lactating ewes in spring and growing lambs in summer. However, the needs for adequate nutrition at mating in autumn and during pregnancy in winter correspond with periods of low or zero growth and must be met by the provision of saved pasture ('foggage'), from conserved grass (hay or silage) or supplementary feeds. It is total grass production which eventually limits the stock-carrying capacity of an area, but grazing control (grazing fields at times of highest demand and resting them at other times) can help to utilize this production more efficiently.

The grass growth curve

The seasonal pattern of grass growth in the UK is characterized in Fig. 2.1. A peak of production is reached in early summer (May), followed by a second, smaller peak in late summer (July–August). Production falls sharply to zero in the winter months (November–March). There are also differences between years both in total yield and in the pattern of production. This calls for some flexibility in grazing systems to allow changes in management to compensate for differences in grass supply.

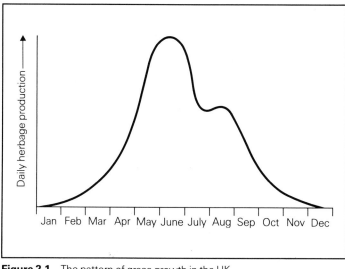

Figure 2.1 The pattern of grass growth in the UK

Factors affecting grass production

Altitude and aspect

Because grass growth rate is largely determined by temperature, production is less, starts later in spring and finishes earlier in autumn at higher altitudes (see Table 2.1).

Table 2.1 The effect of altitude on the start and duration of the growing season (data from a survey conducted at the East of Scotland College of Agriculture)

Site	Altitude (m)	Average start of the growing season	Duration of growth (days)
Kelso, Roxburgh	59	1 April	224
Wolflee, Roxburgh	163	11 April	206
West Linton, Peebleshire	233	20 April	188
Upland site, Peebleshire	450	3 May	162

Grass will also have a shorter growing season on a north-facing slope compared to one that is south facing. Grass production is thus most limited on hill farms, particularly those in less favoured situations. Low mineral status, acidity and waterlogging also contribute to the very much lower productivity of hill pasture.

Species

Different grasses vary in their earliness of growth and flowering and in their overall productivity. Earlier varieties are more productive but suffer a decline in nutritive value after flowering so that the growth of later maturing varieties

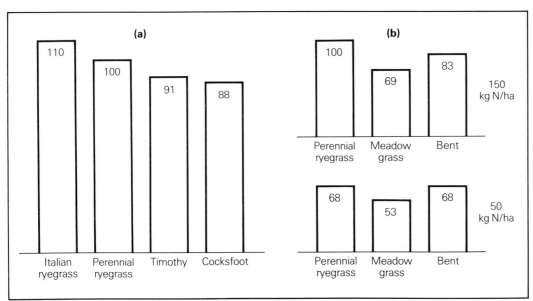

Figure 2.2 Relative yields of different grasses (a) Sown species with 350 kgN /ha (8 cuts) (perennial ryegrass = 100) (b) Weed grasses compared to perennial ryegrass at 150 kgN/ha and 50 kgN/ha (5 cuts) (perennial ryegrass at 150 kgN /ha = 100) (a) From G. Swift (unpublished data) (b) From A. H. Charles, J. L. Jones, M. S. Thornton and T. A. Thomas (1978) 'Changes in sward composition and productivity', *Proceedings of the 10th Occasional Symposium of the British Grassland Society*

coincides better with the peak of demand of the sheep flock.

Species vary in total production with the ryegrasses being most productive (Fig. 2.2). Italian ryegrass (*Lolium multiflorum*) is the earliest maturing and also the highest yielding grass. Different varieties of perennial ryegrass (*L. perenne*) are classified as 'early', 'medium' and 'late'. Both timothy (*Phleum pratense*) and cocksfoot (*Dactylis glomerata*) are less productive, but may be included in a grass mixture for reasons other than total yield, for example, winter hardiness (timothy) and drought resistance (cocksfoot).

Meadow grass (*Poa* spp.), fescues (*Festuca* spp.), bents (*Agrostis* spp.), Yorkshire Fog (*Holcus lanatus*) and the very wiry hill grasses such as mat-grass (*Nardus stricta*) and purple moor grass (*Molinia caerulea*) are far less productive than sown species and respond much less to fertilizer nitrogen. Far less production is therefore possible from rough grazing and hill pastures. Pasture improvement entails the introduction of more productive grasses (mainly ryegrass) and clovers. Factors other than total yield (seasonality, palatability and persistence) affect grass productivity under the given conditions and proposed management. Nor is it always essential to establish a sward of maximum grass productivity. Partial improvement may be sufficient to allow increases in animal production, whereas often the potential of the grass sward can never be realized in terms of ultimate animal product.

Fertilizer

The main components of fertilizer are nitrogen (N), phosphorus (P_2O_5) and potassium (K_2O). There is a linear response in total yield of grass up to a total of 300 kg N/ha, applied regularly throughout the season (see Fig. 2.3). On lowland and inbye fields, an early application is important to ensure good growth of grass in spring and it should be timed to allow uptake by the grass to coincide with the temperature reaching the level required for growth (6 °C for ryegrass). It is normal practice to apply a reasonably large quantity of fertilizer at the first application. On lowground pasture up to

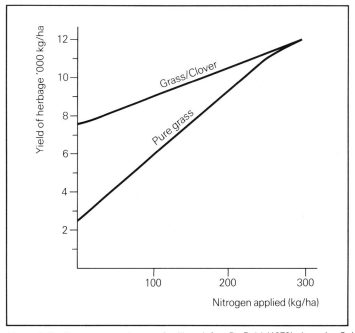

Figure 2.3 Response to nitrogen fertilizer (after D. Reid (1979) *J. agric. Sci., Camb.* **79,** 291-307

100 kg/ha may be applied at the first application and this may be followed by a second application 6 weeks later of about 50 kg N. Successive applications of up to 50 kg N may be given at monthly intervals as required but in the UK an economic response will be obtained only up to mid-September (mid-August in N. England and Scotland). Total quantity of N fertilizer can be adjusted according to stocking rate; experience has shown that the optimum rate for lowground sheep systems is about 12 kg/ewe. Thus, at a stocking rate of 12 ewes/ha, 150 kg N should be applied and this may be divided into one application of 70 kg and two later applications of 40 kg N/ha.

On hill pastures, nitrogen application is usually impossible due to inaccessibility and terrain. When hill and upland pastures are reseeded, 80 kg/ha each of N, P_2O_5 and K_2O (in a compound fertilizer) are valuable for establishment and some regular application of 40 kg N/ha may be possible. However, greater reliance may be placed on clover in hill swards to supply the necessary nitrogen for a moderate level of productivity which exceeds that of the native sward.

The requirements for phosphate and potash vary with soil type and use. When pasture is grazed, most of the phosphate and potash is recycled through the animals and returned to the pasture in dung and urine. When grass is cut for hay or silage, phosphate and potash reserves are depleted and must be restored by fertilizer application. The soil status can be checked by chemical analysis and slag or superphosphate applied in autumn/winter or compound fertilizers used in summer if required.

In order to avoid the risk of 'staggers' (hypomagnesaemic tetany) in lactating ewes compound fertilizers should not be used in the spring; potassium interacts with magnesium in the soil and reduces its availability. Nitrogen should be applied alone at the first application and later applications given as a compound. The application of 50 kg/ha of P_2O_5 and K_2O in the season is sufficient on most lowground grazing land.

Many hill and upland soils are low in phosphate, and reseeding of hill pastures will normally require the prior application of about 150 kg P_2O_5/ha in the form of slag or superphosphate (in addition to the use of a compound fertilizer) to correct this deficiency. Calcium as lime is also needed to correct the acidity of many hill soils. The quantities of lime required can be quite large (5 tonnes or more per hectare) but, as plant growth is limited to the top few centimetres, applications of smaller quantities (2–3 tonnes) may suffice if cultivations are limited to surface work. Further applications of lime will be required after 4–5 years.

Other elements are missing in certain soils. These deficiencies do not affect grass growth, but cause certain specific deficiencies in the stock. These are the so-called 'trace elements' and the common requirements in certain areas are for cobalt (Co) and copper (Cu). Severe Co deficiency causes a serious loss of condition (pining) in ewes and lambs, and chronic deficiency is characterized by unthrifty lambs in late summer. Cobalt deficiency can be corrected by application of cobalt sulphate to pasture, preferably in winter when the grass is short, at the rate of 6 kg/ha to a quarter of the area ($1\frac{1}{2}$ kg/ha overall). Recent increases in the price of cobalt have favoured the alternative method of giving cobalt oxide bullets. These are given by mouth to the sheep and lodge in the rumen or reticulum where they release minute amounts of cobalt for several years. They are not suitable for young lambs where the rumen is not well developed and these may be given short-term treatment by injection of vitamin B_{12}; cobalt is required by the rumen micro-organisms for the synthesis of this vitamin which is in turn absorbed by the animal from its gut.

Copper deficiency causes swayback (neonatal ataxia), a nervous disorder of new-born lambs arising from a deficiency of Cu in the diet of the pregnant ewe. Growing lambs can also be affected, causing reduced liveweight gain. Because there is only a small difference between the amount required and toxic levels for sheep, copper deficiency is best corrected by administering a measured dose of copper sulphate solution orally or by injection of an organic copper complex. Ewes are normally treated in mid-pregnancy on farms with a history of swayback.

Age of sward

Young swards, sown down with some early perennial and/or Italian ryegrass in the mixture, will produce more growth in the early season.

Because the early varieties are less persistent and Italian ryegrass is particularly susceptible to 'winter kill', these varieties disappear as the sward becomes older, and older swards become later in growth and maturity. Later varieties are also less productive, and total production falls with the age of the sward. However, older pasture is denser and less subject to physical damage when wet. It is also deeper rooting and better able to withstand drought conditions in a dry summer. Old permanent pasture can stand quite heavy grazing by sheep and, at moderate levels of nitrogen application, is quite as productive in terms of lamb performance as a young sward.

Clover

Clover makes an important contribution to grass production by virtue of the nitrogen fixation of its root-nodule bacteria. At low nitrogen levels, the productivity of a grass/clover mixture is much higher than that of a pure grass sward (Fig. 2.3) but a mixture responds less to fertilizer nitrogen. Nitrogen may have a detrimental effect on the clover, mainly because of increased competition from grasses, particularly when grazing is lax. Clover is particularly valuable on hill and upland pastures where regular fertilizer applications are difficult or uneconomic. Hill soils, rich in organic matter, contain large quantities of nitrogen in unavailable form. A more available form of nitrogen must be provided and clover provides this through the fixation of atmospheric nitrogen.

White clover is mainly used in the UK. This herbage is high in digestible energy, protein, minerals and vitamins compared to grass, so that it is a particularly good feed, especially for fattening lambs. Red clover is less productive and some concern exists over possible oestrogenic effects of both red and subterranean clover which are known to affect ewe fertility in Australia.

Nitrogen fixation occurs through bacteria (*Rhizobia*) which form nodules on the roots of the plants. Some strains of *Rhizobium* are more active than others and these differ with different varieties of clover. Larger nodules found on the main roots of the plant are the 'effective' *Rhizobia*, and where these are not present the clover is making little contribution to fertility. Trials have been carried out in Australia, New Zealand and recently in the UK on the possibility of 'inoculating' clover seed with effective *Rhizobia*. Results in the UK suggest that this has little lasting effect except on very poor, peaty soils where few *Rhizobia* are found. Little difference has been demonstrated on mineral soils between 'inoculated' and 'uninoculated' clover.

There are wide-ranging estimates on the nitrogen contributions of clover to mixed swards (ranging from 90 to 200 kg/ha), but the contribution can be considerably less when establishment or persistence of the clover is poor. An example of the contribution of clover to herbage nitrogen yield is shown in Fig. 2.4.

Reliance on clover requires good establishment and correct management, neither of which are easy to ensure. One problem is that grass mixtures are normally drilled which ensures an even 'take' of grass, but this tends to bury the clover seed too deeply and broadcasting is a better method for clover establishment. The ideal method is to drill the grass and broadcast the clover, and although machines have been developed to combine these operations, it is not a normal practice.

Close grazing is essential to maintain clover and prevent 'shading out' by the grass. Clover can withstand up to 100 kg N in one application without detriment, provided that the resulting grass production is fully utilized by grazing livestock and the sward kept short.

Grass mixtures

By sowing a mixture of grass and clover varieties, it is possible to obtain a sward of high productivity throughout the grazing season, which is adapted to the particular conditions and which will withstand grazing and/or cutting for a number of years.

Although Italian ryegrass is the most productive and earliest grass, it is subject to 'winter kill' and rarely lasts more than one or two seasons. In medium-term leys, a few kilos of

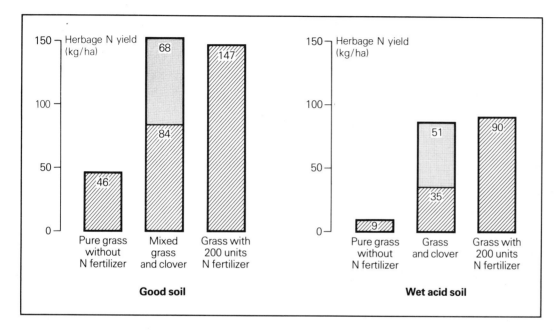

Figure 2.4 Clover contribution to herbage nitrogen production of mixed swards on hill soils. (From P. Newbould and A. Haystead (1977) Hill Farming Research Organisation Seventh Report (after J. M. M. Munro and D. A. Davies (1974) *J. Br. Grassld. Soc.* **29**, 213–23)

Italian ryegrass are usually included for the benefits of first-year production. Perennial ryegrasses form the bulk of most modern seed mixtures and a combination of early, medium and late varieties provides for a balanced pattern of grass supply. Some early varieties are desirable for swards that are to be cut, but for grazing purposes emphasis should be placed on the medium–late varieties. Timothy is highly palatable and suitable for less fertile or wet conditions. It is very winter hardy but does not withstand heavy stocking. Cocksfoot is tolerant of low fertility and resistant to drought and is particularly favoured in low-rainfall areas. However, it is not very palatable and often becomes rank in mixed swards that are not heavily grazed. Other grasses, such as red fescue, rough and smooth stalked meadow grass, have traditionally been included at low rates as an 'insurance' against failure of other grasses, but they have few production attributes except possibly in hill reseeds where fertility is fairly low.

White clover usually represents about 6–8 per cent of the seed sown and the favoured varieties are the medium-leaved types (Huia, S100) and wild white (Kent, A184). Examples of suitable grass mixtures for lowground leys and hill reseeds are given in Table 2.2.

Grazing management

The most effective utilization of pasture is obtained by controlling grazing so that the seasonal production is used by the sheep in relation to its needs. The growth of the pasture is also improved by avoiding grazing in late winter and early spring, grazing heavily in summer (to maintain vegetative growth and delay flowering) and moderate defoliation in autumn (to prevent frost damage to leafy material).

The peak demands for the sheep flock are around and during mating, late pregnancy and lactation. These correspond with autumn, late winter/early spring and summer. It is not possible to satisfy these needs from grass

Table 2.2 Grass mixtures for sheep grazing

1. *Medium or long ley, without timothy (light medium, loam)*

4 kg	Italian ryegrass
6 kg	Early perennial ryegrass
8 kg	Medium late perennial ryegrass
6 kg	Medium late perennial ryegrass (tetraploid)
7 kg	Late perennial ryegrass
2 kg	White clover*
33 kg	per hectare, broadcast

2. *Medium or long ley, with timothy (medium heavy, loam)*

4 kg	Italian ryegrass
6 kg	Early perennial ryegrass
8 kg	Medium late perennial ryegrass
7 kg	Late perennial ryegrass
4 kg	Early timothy
2 kg	Intermediate timothy
2 kg	White clover*
33 kg	per hectare, broadcast

3. *Hill reseed*

6 kg	Early perennial ryegrass
6 kg	Medium late perennial ryegrass
9 kg	Late perennial ryegrass
3 kg	Early timothy
3 kg	Intermediate or late timothy
3 kg	Early cocksfoot
2 kg	Late cocksfoot
4 kg	Creeping red fescue

Table 2.2 continued

1½ kg	White clover
½ kg	Wild white clover
38 kg	per hectare

*Substitute 0.5 kg wild white clover if for longer than 5 years.

alone because the winter/spring period coincides with a time when there is little growth and a danger of pasture damage.

Wintering systems

Supplementary feeding of cereals, forage crops and conserved grass are required during the winter period. This can be done on a small area which is not required for spring grazing (a hayfield or a permanent pasture 'sacrifice' area). There is evidence that stocking rates of up to 35 ewes/ha cause little long-term damage to well-drained pasture and total yield of grass is hardly affected, although spring growth may be somewhat delayed.

On wet areas, where heavy winter stocking rates can lead to physical damage of pasture, inwintering of the flock may be appropriate. However, capital costs of buildings and interest on this capital, together with added costs of feeding sheep indoors, must be set against the advantages obtained.

Inwintering is justified only where outwintering is virtually impossible owing to very wet soil and adverse climatic conditions.

The traditional practice of allowing sheep to graze all over the farm in winter delays spring growth on all the fields and has little advantage to the ewes; the requirement for supplementary feed is nearly the same whatever the grazing allowance because winter grass makes only a small contribution to the diet.

Early spring management

After lambing there is a high demand for food by the lactating ewe, and this does not always coincide with onset of spring growth. Lambing should be as near to the onset of grass growth as possible, which means later in the north and on higher farms. On lowland farms late lambing means that the lambs are too small to make use of the peak of growth in May and June and that they are getting large in July and August when growth begins to decline. Earlier lambing makes better use of the pattern of grass growth and means that some lambs can be sold fat before grass growth declines (Fig. 2.5).

The problem of a shortage of feed in early lactation, that occurs when ewes lamb in early spring, can be overcome by supplementary feeding at pasture either with concentrates or, more cheaply, with fodder crops such as swedes.

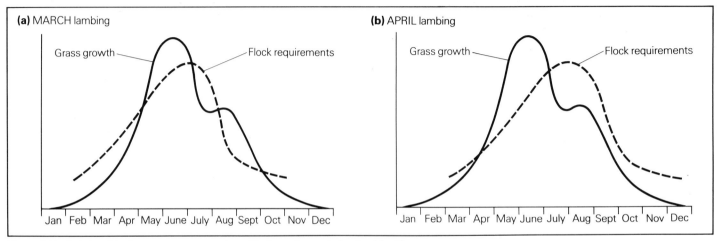

(a) MARCH lambing

Grass growth

Flock requirements

Jan Feb Mar Apr May June July Aug Sept Oct Nov Dec

(b) APRIL lambing

Grass growth

Flock requirements

Jan Feb Mar Apr May June July Aug Sept Oct Nov Dec

Figure 2.5 Relationships between grass growth and flock requirements (lowland flock)

On many farms sheep are kept off their summer grazing area in spring in order to encourage subsequent grass growth. Fields intended for conservation or cattle grazing are grazed first and sheep moved onto their summer grazing when growth begins. Although this practice is beneficial to sheep grazing fields, it results in delayed cutting of hay and later turnout of cattle and there is little evidence to suggest that sheep performance is improved. In an experiment at Edinburgh, on dry, well-fertilized land, neither sheep performance nor subsequent grass production were affected by stocking up the sheep fields in early spring and supplementing the ewes with cereals and roots. Cattle turnout was earlier (with savings in winter feed) and hay made sooner.

Problems of physical damage of pasture in spring may occur in wet areas. In practice, a compromise is the best solution with some early grazing of hayfields before fully stocking up the sheep fields. There will be little effect on hay yield or quality if it is given adequate fertilizer. Young grass fields that are to be stocked heavily in summer will benefit from light grazing only in spring. Hayfields should be shut up as soon as possible when grass growth begins. Silage and cattle grazing fields should not be grazed by sheep in spring.

Summer management

The decision to be made in summer grazing is whether to employ set-stocking (where the sheep remain on the same fields at a given overall stocking rate) or a system of rotational grazing where fields or paddocks are used in succession (at a much higher stocking rate on each) and then rested to allow regrowth.

In the past, it was claimed that better use of pasture production could be achieved with rotational grazing. Paddock grazing for sheep has usually involved six to eight paddocks. A further refinement was the introduction of a creep arrangement for the lambs (allowing them to graze a rested paddock ahead of the ewes). This 'forward creep grazing system' was developed at Newcastle in the 1950s by Professor M. McG Cooper and his colleagues. The main features of the system are shown in Fig. 2.6. Wheeler (1962)* has discussed the results of experiments comparing rotational grazing with set-stocking (mainly cattle). One claim made for

*J. L. Wheeler, *Herbage Abstracts*, 1962, **32**, 1–7.

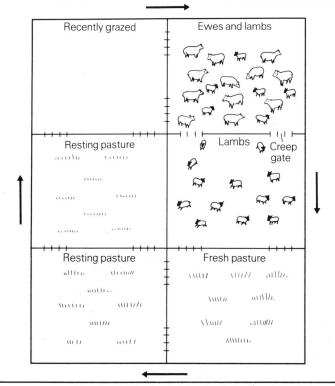

Figure 2.6 The forward creep grazing system

the former is that total yield of grass is increased and this is based on early experiments at Cambridge in the 1930s. Lucerne, cocksfoot and ryegrasses were used in these trials and close cut. These particular grasses require a long period of recovery to restore carbohydrate reserves after grazing or cutting. However, clovers and other grasses recover much more quickly and the benefits of allowing recovery are less. Nevertheless, there is still more complete defoliation with rotational grazing because continuous grazing permits selectivity.

The other claim was that rotational grazing gave better worm control. This was based on the observation of a 3-week life cycle in parasitic roundworms. It is now known that the life cycles are highly variable over the season and the only effective control method is a clean grazing system (see Ch. 3). The main advantage of paddock grazing is that it allows adjustments to be made in the rate of grass use over the season and for paddocks to be set aside for conservation.

In fact, the needs of the sheep flock in summer can fit well with the pattern of grass growth (see Fig. 2.5) and set-stocking is a simpler system which requires less expertise to succeed.

If a paddock grazing system is managed incorrectly so that sheep are allowed to graze too long in one paddock, they will become short of feed and performance will suffer. This cannot happen with set-stocking unless total grass production is insufficient and this point will be reached regardless of the grazing system if the overall stocking rate is too high.

In all systems, nitrogen must be applied regularly to maintain grass supply. There is no evidence that applying moderate levels of nitrogen among grazing sheep does them any harm.

Weaning

Considerable variation in grass supply occurs in late summer. If grass is in short supply the requirements of ewes and lambs may not be met. If this is the case, the ewes can be removed from the system, ideally to hill or rough grazing, or heavily stocked on one or two fields, leaving the rest of the grass for the lambs. If grass is in surplus, ewes can be left to control it. Grass quality must be maintained at high digestibility by keeping the grass short and leafy and undergrazing is as undesirable as overgrazing. Time of weaning can be varied to control the grass and optimize ewe and lamb nutrition.

As soon as aftermaths have made reasonable growth (50–60 mm) after hay- or silage-making, lambs can be moved onto this area.

Autumn management

Ewes require a good grass supply before and during mating. Late application of fertilizer and resting of pasture after weaning (by keeping ewes on rough grazing or at a high stocking rate on one field) allows grass to freshen up for this period. If ewes are in poor body condition, they must be returned to good pasture well before mating.

In mixed farms, the ewes can make use of arable stubbles, new-sown leys, aftermath remaining after the lambs have been sold or the fields vacated by cattle after housing. If grass is in short supply and ewe body condition poor, supplementary feeding of concentrates on forage crops may be needed.

The farm system

Planning the grazing system requires consideration of many factors. The sheep flock must be integrated with other farm enterprises (crops, cattle and conservation). Grass may be used alternately for sheep, cattle and for conservation which is beneficial to long-term sward maintenance and permits effective parasite control.

Lowland and upland sheep production

3

The potential performance of sheep on sown pastures and good permanent grass in lowground and upland areas is far greater than in the hills. However, the higher fixed costs (particularly of land) and the possibility of profitable alternatives (crops or other livestock enterprises) make it important that output from the sheep enterprise is competitive. Sheep on hill farms may have a low productivity level but they use resources of limited value; the targets for sheep production in the lowlands must be set much higher to justify their inclusion in a mixed farming system.

There are a variety of situations into which a sheep flock may fit. On an arable farm, sheep utilize the grass break in a mainly cereal rotation and unsaleable by-products such as stubbles and sugar-beet tops, and also any unploughable grass. Because sheep are a secondary enterprise (and invariably less profitable than crop production), the aim must be to maximize crop production and to confine sheep to the minimum area necessary to provide a suitable fertility break. Increasing the stocking rate means keeping the same number of sheep on a smaller area.

In livestock rearing and upland areas, sheep and cattle are usually the main enterprises because of the limitations to crop production of the climate, soil and topography. Here again the emphasis must be on maximum output per hectare as this will be the major determinant of whole farm output. A higher stocking rate means keeping more sheep (and cattle) on the same area.

On all farms, the sheep must be integrated with the other enterprises on the farm to their mutual advantage. This means planning a rotation that ensures the maximum number of years of crop production in relation to the suitability of the land, with a grass break that is used with maximum efficiency by the appropriate balance of sheep and cattle.

Limitations on stocking rate

Traditionally, the relative stocking rate of sheep has been far lower than that for beef or dairy cows. The average use of nitrogen fertilizer for sheep grazing on MLC-recorded farms is 90 kg/ha on rotational grass and 62 kg/ha on permanent grass, whereas the use of over 300 kg N/ha is not uncommon on intensive dairy farms. Pasture production is thus rarely the limiting factor in sheep grazing. Attempts to increase nitrogen usage and stocking rate in the past have been disappointing because increased sheep numbers have resulted in reduced lamb performance and the expected increase in overall output was not achieved.

The main reason is the increased incidence of gastro-intestinal worm parasites associated with high stocking rates of sheep year after year on the same pasture. The susceptibility of lambs to worms leads to chronic parasitism which reduces lamb growth rates and spoils their appearance. Acute parasitism, causing wasting and death, may occur at extreme levels. Anthelmintic (worm-killing) drugs are only a partial solution because their effectiveness is limited when lambs are returned to infected pasture immediately after dosing.

The only satisfactory solution is to graze ewes and lambs on clean pasture each year, that is, pasture that has not been infected by worms in the previous year.

The worm problem

Much of our current understanding of sheep worms stems from the studies conducted by Dr R. J. Thomas at the University of Newcastle over the last 20 years. There are two main sources of worm infestation (see Fig. 3.1). Lambs and other young sheep become infected by eating worm larvae on the pasture and multiply the infection by depositing eggs with their faeces. The eggs of certain species of worm (mainly *Nematodirus* and *Ostertagia* spp.) overwinter on pasture and reinfect lambs in the following spring. Thus, the levels and effects of worm parasites increase if lambs are grazed on the same grass year after year.

Ewes are resistant to worms for most of the year, but their resistance breaks down in spring and worm-egg output increases rapidly after lambing (the 'post-lambing rise'). These eggs develop into larvae on the herbage and become a second source of infection for lambs later in the season.

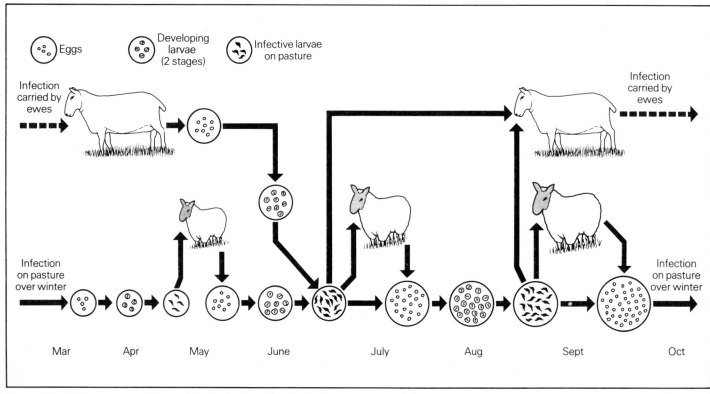

Figure 3.1 The two sources of worm infection and how they work

Clean pasture

Overwintered worm eggs are present on pasture grazed by lambs or young sheep in the previous year. Clean pasture is grass that has not been grazed by lambs or young sheep for at least 12 months. It may be used for cattle or conservation in summer and by weaned calves or dosed ewes in autumn. Young grass (sown the previous year under barley) will also be clean provided it is not grazed by lambs after harvest. By the simple expedient of grazing lambs on clean grass, overwintered worm infestation disappears.

The problem of contamination of pasture by the ewes after lambing is avoided if ewes are dosed with an anthelmintic as they go onto the clean pasture (Fig. 3.2). Lambs will not be infected at this time provided they have not started to graze,

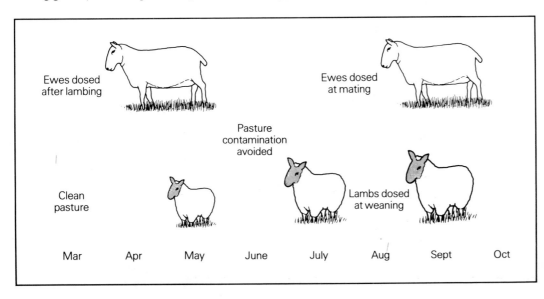

Figure 3.2 How clean pasture and correct dosing avoids the worm problem

that is, they are less than 3 weeks of age. Ewes and any lambs over 3 weeks old require dosing as they move from the lambing field onto the clean summer grazing. These practices will keep worm infestation to a very low level and it is normally not necessary to dose lambs during the summer.

At weaning, lambs are dosed and moved onto hay or silage aftermaths which will also be clean because worm eggs hatch in spring and no lambs are present to multiply the infection. In autumn, feeder lambs and young female replacements are kept off fields required for next year's summer grazing of ewes and lambs. These are used for cattle or dosed ewes as already described; ewes do not contaminate pasture in autumn because they are resistant to worms at this time, but a single anthelmintic dose provides extra insurance.

A sheep system on an arable farm

The provision of clean grass in the mainly arable farm is less difficult than in a mainly grass situation. As the aim is to maximize cereal production, a short-term ley with a high stocking density is required. First-year grass is automatically clean provided it was not used for lambs after harvest in the sowing year. The second-year grass will be required for production of winter feed (hay or silage) for the ewes and will provide clean aftermath for lambs after weaning. Because the hay area required for sheep is much less than the summer grazing area some grass can be ploughed after only one year. This may seem a costly process, given the establishment costs of a new ley, but the costs can be set against the production that will be obtained from the extra cereal crop grown. Alternatively, the second-year grass not required for hay may be grazed by fattening cattle.

The system (March lambing)

1. Ewes run on stubbles after weaning until early September.
2. Ewes graze young grass before mating and utilize all the remaining grass area during and after mating. (In southern areas, where new grass is available early because of early harvest, draft ewes may be fattened on it, or mating advanced to utilize this grass.) Ewes are dosed with anthelmintic before grazing next year's clean fields. One ram is required for every 40 ewes mating. The rams are withdrawn after 5–6 weeks to avoid a protracted lambing.
3. From January onwards ewes are concentrated on grass about to be ploughed up for barley. All remaining grass is rested to allow good spring growth.
4. Lambing takes place either in a small permanent lambing field or next year's hay ground (second-year grass); the second alternative avoids the build-up of disease which may occur in a lambing field that is used every year. Ewes and lambs are moved on to the young grass as soon as possible (ewes dosed on entry). Lambs born at the start of the

lambing season are moved on first and the field stocked up gradually over 3–4 weeks.

5. Once the target stocking rate is reached, grass growth will be well established and ewes and lambs may be set stocked until weaning. Lambs are sold fat as soon as possible.

6. At weaning, lambs are dosed and moved on to hay aftermaths. Ewes graze remaining grass and stubbles when available.

An integrated beef and sheep system for upland areas

On longer-term leys and permanent pasture, clean grass may be provided by an integrated beef, sheep and conservation system. 'Clean grass' for sheep is achieved by a 3-year rotation of: sheep; hay; cattle; in that order. The system was developed 6 years ago by Mr W. Rutter of the East of Scotland College of Agriculture at House o' Muir Farm, near Edinburgh. The farm consists of productive pasture at 250–300 m above sea-level and is divided into three roughly equal areas with one-third for sheep, one-third for conservation and one-third for cattle. (Fig. 3.3).

The main points of the management are as follows:

1. Ewes lamb in March and early April. They are dosed after lambing together with any lambs over 3 weeks old as they go on to the clean field (A).

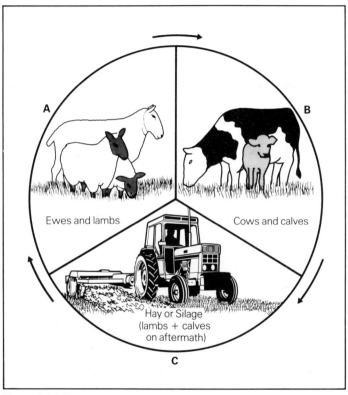

Figure 3.3   The clean grazing system at House o' Muir

Figure 3.4 Ewes at mating are dosed if grazing next year's clean grass

2. Ewes and lambs are set-stocked from April to July on area A which was grazed by cattle in the previous year. Cows and calves graze area B; area C is cut for hay.
3. In late July/early August, lambs that have not been sold fat are dosed and moved on to the hay aftermath (C) together with weaned calves. Ewes remain on area A, cows on area B.
4. Cattle are housed in the autumn and lambs sold or moved to forage crops, allowing the ewes to graze A, B and C as required to provide the best grass before, during and after mating; the ewes do not contaminate next year's clean grass (B) because they are resistant to worms at this time, but are dosed in any case as an extra insurance (Fig. 3.4).
5. From January onwards, ewes are confined to half of area A and the remaining grass rested to permit good spring growth.
6. After lambing, ewes and young lambs move onto the other half of area A for a short time and then are dosed and moved onto the new clean field (B) which was last year's cattle area. Cattle then graze C and hay is made on area A.

At the College farm, the sheep stocking rate is 17.5 Halfbred ewes and 30 lambs/ha. A total of 210 kg N/ha is applied in doses of 100 kg in March, 50 kg in May, 35 kg in June and 25 kg in July. Lamb growth rates have averaged 270 g/day from birth to weaning (at 120 days) over 5 years and 50–60 per cent of lambs have been sold fat by weaning. The stock have remained uniformly clean and healthy, despite the fact that lambs are not dosed for worms until weaning.

Cattle have been stocked at 3.5 cows and their calves per hectare. Thus, 25.5 ha of grass produces 250 lambs, 30 suckled calves and 50 tonnes of hay each year.

Alternating sheep and cattle

On steeper upland farms, it is not possible to cut every field for hay or silage. Another method of providing clean grass is to alternate sheep and cattle on an annual basis. Hay is made on mowable fields elsewhere on the farm (Fig. 3.5).

The management system is the same as that described on pages 27–28, except that ewes are wintered either on part of area C or on a separate area of rough grazing. In the next year, sheep go onto the new clean field (B) and the cattle graze A while hay or silage is again made on field C. This system has also been operated by the East of Scotland College at Woodhouselee Farm with 216 Greyface ewes grazing 17 ha of upland grass, alternated annually with 17 ha used for cattle. Silage has been made on short-term leys on the arable ground below and this has provided aftermaths for lambs and calves. Lamb performance over 3 years averaged 301 g/day from birth to weaning (at 120 days) and no anthelmintic doses have been given to lambs before weaning.

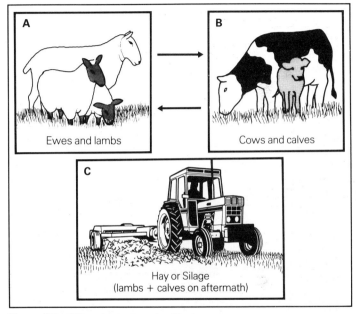

Figure 3.5   Alternate sheep and cattle

Stocking rate on clean grass

Because worm parasites are effectively controlled when sheep graze clean grass, the stocking rate can be increased above the traditionally accepted levels without depressing lamb performance. The number of sheep per hectare is not limited by worms and can be related better to grass production and fertilizer use. One guide to workable stocking rates for summer grazing is the potential barley yield of the land (Table 3.1). Land capable of growing 5 tonnes of barley per hectare is capable of supporting 20 ewes and their lambs per hectare from lambing to weaning. On poorer land, capable of growing $3\frac{1}{2}$ tonnes of barley, as is common in marginal areas, 14 ewes/ha is likely to be a more realistic target.

Table 3.1 Stocking rate of sheep in relation to potential barley yield (after W. Rutter (1975) *Sheep from Grass*, East of Scotland College of Agriculture, Bulletin 13)

	Potential barley yield (tonnes/ha)					
	$2\frac{1}{2}$	3	$3\frac{1}{2}$	4	$4\frac{1}{2}$	5
Summer grazing capability (ewes/ha)	10	12	14	16	18	20

Fertilizer application must be adjusted to suit the conditions and the intended level of stocking. Annual nitrogen requirement is approximately 12 kg/ewe grazing. The timing of application should be aimed at providing good spring growth and a continuing supply of grass throughout the grazing season. A suitable nitrogen fertilizer programme for three stocking rates are given in Table 3.2. These ought to be adjusted to suit local conditions of temperature, rainfall and soil conditions. The requirements for phosphate and potash will depend on soil analysis and can be met either by application of slag, superphosphate or potassium superphosphate in winter, or by use of a compound in summer. Nitrogen alone should be applied at the first application because potash may reduce the availability of magnesium and cause hypomagnesaemia ('staggers') in ewes.

Table 3.2 Nitrogen fertilizer programme

	kg N/ewe	kg N/ha		
		14 ewes/ha	16 ewes/ha	18 ewes/ha
Late March	6	80	90	100
Mid-May	3	40	45	50
Mid-late June	3	40	45	50
Total	12	160	180	200

Fitting sheep into the farming system

A rotation is planned to include the desired area for sheep, cattle, conservation, cereals and roots. The most appropriate rotation for the farm must also take into account such factors as soil type, topography, field size and location.

The farm is first divided into different classes of land, the simplest division being into arable (including temporary grass) and permanent pasture. Within each class of land, the fields are grouped into the most convenient 'breaks' to give a

Figure 3.6   Seventeen and a half ewes per ha. on clean grass with 200 kg N/ha

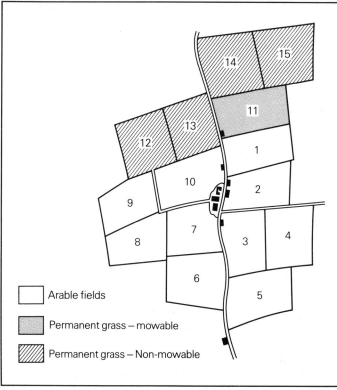

Arable fields

Permanent grass – mowable

Permanent grass – Non-mowable

Figure 3.7   Example of a farm plan

rotation of reasonable length. The rotation on the grass fields ensures that sheep always follow cattle, undersown cereals or aftermath grazed by calves. Permanent grass fields which cannot be cut alternate between sheep and cattle.

An example of a simple farm plan is shown in Fig. 3.7 and the rotation plan for the farm for the next 5 years is given in Table 3.3. For simplicity, all the fields are the same size (10 ha). The rotation involves 5 years' cropping and 5 years' grass on each arable field and the non-mowable permanent grass fields are alternated between sheep and cattle. A permanent grass field near the shepherd's house is used as the lambing field and then cut for hay every year. Thus the 150 ha farm grows 40 ha barley, 10 ha turnips and 30 ha hay, and there are 40 ha of grass for sheep and 30 ha for cattle. With the use of 170 kg N/ha, this grass can support 15 ewes/ha or 3 cows/ha giving a total of 600 ewes and 90 cows.

Introducing the clean grazing system onto a mixed farm

If the farm is currently not working a clean grazing policy, it is likely that the sheep will have grazed all the grass fields each year. To introduce clean grass, the young grass field should be kept clean by avoiding the use of the undersown stubbles for lambs. This field can therefore be stocked at 15 ewes and lambs per hectare in the next year and this means that another

Table 3.3 5-year rotation plan

Arable fields

Field	ha	1980	1981	1982	1983	1984
1	10	Barley	Barley	Turnips	Barley	Barley*
2	10	Barley	Turnips	Barley	Barley*	Sheep
3	10	Turnips	Barley	Barley*	Sheep	Hay
4	10	Barley	Barley*	Sheep	Hay	Cattle
5	10	Barley*	Sheep	Hay	Cattle	Sheep
6	10	Sheep	Hay	Cattle	Sheep	Hay
7	10	Hay	Cattle	Sheep	Hay	Barley
8	10	Cattle	Sheep	Hay	Barley	Barley
9	10	Sheep	Hay	Barley	Barley	Turnips
10	10	Hay	Barley	Barley	Turnips	Barley

Permanent grass

Field	ha	1980	1981	1982	1983	1984
11	10	Hay	Hay	Hay	Hay	Hay
12	10	Sheep	Cattle	Sheep	Cattle	Sheep
13	10	Sheep	Cattle	Sheep	Cattle	Sheep
14	10	Cattle	Sheep	Cattle	Sheep	Cattle
15	10	Cattle	Sheep	Cattle	Sheep	Cattle

*Barley undersown with grass for 5-year ley.

field can be stocked exclusively with cattle. In the second year, there will be two clean fields – the young grass and the field grazed by cattle in the previous year.

On permanent grass fields, it is necessary to graze the sheep for one year on dirty grass and to employ a strict routine anthelmintic dosing programme to avoid a worm problem.

The following year, the fields grazed by cattle will provide clean grazing for the sheep.

If possible, the farm should be divided into blocks so that all the sheep are grazed in adjacent fields and all the cattle in adjacent fields. This makes management easier and reduces the dependence on internal fencing. The simplest system on a mainly grass farm is to split the land area into two blocks with sheep at one end and cattle at the other, alternating each year. Hay and forage crops are fitted into the plan with the fields required for grazing lambs (aftermath, rape and swedes) on the 'sheep' side.

Different proportions of sheep and cattle

Equal proportions of sheep and cattle are not an essential feature for the application of clean grazing. Differences can be accommodated by different lengths of grass ley. Examples of different proportions of sheep and cattle on 3- and 4-year leys are shown in Fig. 3.8. The grazing requirements of five ewes are assumed to be similar to those of one cow.

Older cows with suckling calves are less susceptible to worm problems than sheep, but the clean grazing system will benefit young fattening cattle as well as sheep by avoiding the problem of overwintered worm infestation.

The most difficult situation for intensive sheep grazing is when there are all sheep and no cattle (except on the mainly

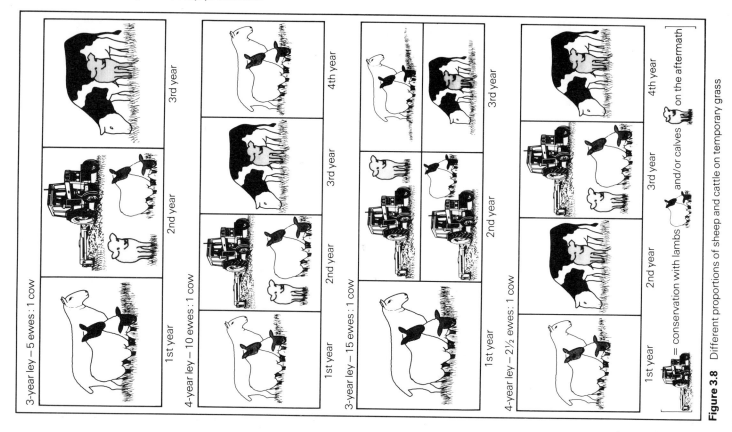

3-year ley – 5 ewes : 1 cow

1st year 2nd year 3rd year

4-year ley – 10 ewes : 1 cow

1st year 2nd year 3rd year 4th year

3-year ley – 15 ewes : 1 cow

1st year 2nd year 3rd year

4-year ley – 2½ ewes : 1 cow

1st year 2nd year 3rd year 4th year

= conservation with lambs and/or calves on the aftermath

Figure 3.8 Different proportions of sheep and cattle on temporary grass

arable farm where the sheep are kept on short-term leys). On longer-term grass, it will be impossible to keep all sheep on clean pasture. Maximum advantage must be taken of any young grass which is introduced, and this should be used for ewes with twins rather than those with singles because twins are more dependent on grazing and therefore more susceptible to worm infection. The level of worm infestation will be fairly low following a year of conservation with careful dosing of lambs onto the aftermath (using a modern anthelmintic). Grazing and conservation can, therefore, be alternated annually to reduce the worm problem.

Sheep on dirty grass

High stocking rates of sheep work best on clean pasture. Where it is impossible to provide clean pasture for all the sheep, it is necessary to resort to more frequent use of anthelmintics. Ewes must be dosed after lambing and lambs dosed in mid-May when the number of *Nematodirus* larvae reaches a peak on pasture. Thereafter the lambs will need to be dosed every 3–4 weeks until weaning when they can be dosed as they move onto hay or silage aftermaths. Mixed grazing with cattle will help to reduce the worm problem, but is not as effective as alternate grazing and clean grass.

Hill sheep production

4

A farm is described as 'hill' when about 90 per cent of the land is rough grazing. The pasture is composed of heath grasses, shrubs or bog plants. Within this category there is tremendous variation in the nature and quality of the soil and its associated vegetation. Although this type of land is usually steep and at high altitudes (above 300 m), some less favoured areas occur at lower levels and have the same low production potential. Hill farms suffer to a varying extent from the limitations of several or all of the following factors:

1. Low temperature;
2. High rainfall;
3. Soil acidity;
4. Impeded drainage;
5. Steep terrain;

The effects of these factors on sheep production are both direct – through the effects of climatic conditions on animal performance and survival, and indirect – through the effects on soil type, nutrient status, vegetation type and pasture production.

The other major limitation on hill farms, resulting from low productivity, is the shortage of capital for investment in improvement. Costs of fencing, cultivations, soil and pasture treatments are all greater in hill farms than on the lowground because of difficulties of access and topography and the extent of the problems involved. It is vital to ensure that the strategy adopted to increase output is cost-effective, providing a sufficient return quickly enough to repay both capital costs and interest charges. Grants may be available from Government sources, but it is still essential to obtain the maximum response from the inputs.

Hill soils and vegetation

Hill soils range from the mineral-rich 'brown earths' to the organic 'peats' (Fig. 4.1). Podsols are leached by rainwater and the minerals have been washed down from the topsoil leaving them more acid than the brown earths. Gleys are soils which are waterlogged for considerable periods of the year. Peats are organic soils, arbitrarily defined as more than 0.3 m deep. Where the peat is shallower, the soil is described as a peaty podsol or peaty gley, depending on the nature of the subsoil.

On the mineral soils, the pasture is composed of bents (*Agrostis* spp.) and fescues (*Festuca* spp.), and clover (*Trifolium repens*) may be found on less acid sites. Such pasture is valuable for grazing but, where it is consistently undergrazed, bracken (*Pteridium aquilinum*) tends to encroach. On more acid and wetter ground, clover is absent and purple moor grass (*Molinia caerulea*) or mat-grass (*Nardus stricta*) are characteristic. These coarser hill grasses are unpalatable and of little value to livestock except in spring and early summer. Where mat-grass, wavy hair grass (*Deschampsia flexuosa*) and rushes (*Juncus* spp.) are the

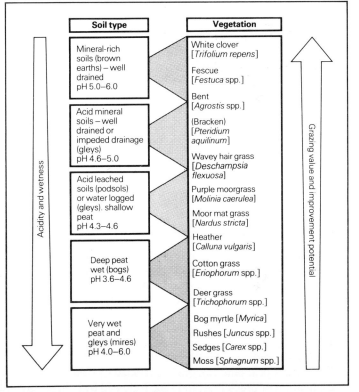

Figure 4.1   Types of hill soil and characteristic vegetation

common species, the soil is very wet and acid.

Heather (*Calluna vulgaris*) is the main vegetation on podsols but, where heather is removed by burning or heavy grazing, purple moor grass and mat grass will appear. On the deep peat soils, heather is again found with cotton grass (*Eriophorum* spp.), deer grass (*Trichophorum* spp.) and moss (*Sphagnum* spp.).

A range of pasture and soil types can be identified on a hill farm and this can be used to indicate the areas of better grazing value and potential for improvement, as shown in Fig. 4.1. They are listed in order of productivity and the aim of land improvement is to convert the poorer vegetation types to one of the better.

Methods of pasture improvement

Grazing control is essential to effective pasture improvement and utilization, so fencing of the area to be improved is the first priority. The method used to improve a given area of land depends partly on the present vegetation type and soil and partly on the objective. It is hardest to improve the poorest soils (wet peaty) and also hard to make the final step to a very productive ryegrass/clover sward. Both these require cultivation and reseeding.

Very wet peat land can only be tackled by drainage and application of much lime and slag, followed by cultivation and seeding. This is a costly operation and would be the last to be tackled in an improvement programme.

Considerable improvement can be made on acid heathland by enclosure, application of lime and slag and increased grazing pressure which can remove the dominant moor and mat grass and make way for bent–fescue swards. Improvement of poorer quality bent–fescue pasture can be achieved by introduction of clover using techniques such as 'slot-seeding', where clover seed is put into a narrow slot cut in the turf by a special machine. This can be used where irregular surface and stones make cultivation impractical. However, the final step is the reseeding of the pasture with productive species – usually a mixture of perennial ryegrass and white clover. The minimum cultivation required to establish a seed-bed is desirable and rotavating of the top few centimetres is successful on mineral soils or shallow peat. This is preceded by application of lime and slag. Finally, grass and clover seed may be drilled or broadcast.

Another method of seed-bed preparation of poor mineral soils is fencing, followed by 'mob-stocking' with cattle to break up the surface and remove existing vegetation.

Pioneer cropping

An alternative is to rotavate and then sow a crop of rape or Dutch turnips in the first year (June and July). Lambs are fattened on the crop and a grass and clover mixture is sown in the following year. A second forage crop may be grown before reseeding with grass and clover and this helps to establish a good soil structure and fertility.

The cash return from fattening lambs repays the majority of the costs of improvement and may even offset the cost of the final pasture establishment. So it is often more economic to grow a pioneer crop and then reseed rather than to attempt partial reclamation by lime/slag and controlled grazing alone.

Choosing the site for improvement

On a hill farm, choosing the best site to begin improvement is vital both in relation to cost, likely success and final utilization of the improved area. Mineral soils and shallow peat associated with the grass heaths are the easiest to improve. Shrub heaths (heather) may be improved relatively easily where the peat is fairly shallow and the land does not require tile draining. Wet land, requiring drainage, demands a large capital input and should be left till later, but many peaty gleys and podsols overlying impeded drainage layers may be improved by deep ploughing or subsoiling and strategic placement of drains to collect run-off water.

The area chosen must be convenient for access and have reasonable shelter as sheep will be fenced onto this area at certain times. It is also important not to enclose all good grazing areas on the hill as these will be needed for wintering, and for single-bearing ewes in lactation and after weaning. A balance must be struck between hill grazing and improved areas to allow an effective system to be practised.

Increasing production from hill farms

Hill farms vary from those with little or no good pasture and enclosed land to farms with considerable improved land offering scope for grazing control and increased lamb production. Any one farm may fall anywhere along this progression and, to improve production, the aim should be to make the necessary improvements to move up the chain towards a better system.

The open hill

On an underdeveloped hill farm there is no enclosure and the sheep range freely over the whole area. Some productive fields near the steading are essential if winter feed is to be made and these will normally be devoted to hay-making and perhaps a field of oats or barley to provide winter feed (Fig. 4.2).

Stocking rate of ewes and lamb production is limited by the native vegetation of the hill and its annual production. Of particular significance is the winter carrying capacity. No

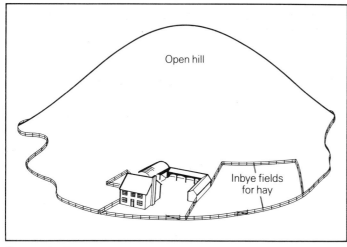

Figure 4.2   The open hill

control is possible at tupping time or lambing, and the sheep are normally brought off the hill only for counting and castrating the lambs in summer, shearing the ewes and weaning the lambs, which must be sold 'store' off the farm. The stocking rate will vary from one ewe per hectare to one ewe per 10 ha, depending on the land. Lambing percentage is rarely better than 80 lambs weaned per 100 ewes with a large proportion of the ewe lambs (at least 50%) required for replacements.

Improving performance on the open hill

Because of the highly seasonal nature of hill pasture production, the lowest level of nutrition is provided in winter when ewes are pregnant. Although it is unlikely that many ewes will produce twins, it is important to ensure that nutrition is adequate for ewes carrying single lambs and to prevent excessive depletion of body reserves. Improvement in the winter nutrition of the ewes is the first step in improving performance. Hay is fed in bad weather and some supplementary cereals, concentrates and/or feed blocks are given in the last weeks of pregnancy and early lactation.

In order to feed effectively, sites are chosen in relation to the normal grazing pattern of the sheep and feed stored in weatherproof containers at these sites for ease of distribution. This system still depends on access to the main grazing areas and this requires the provision of access to the outlying areas. An initial investment may be required to make one or more lengths of track for vehicles to cross the area.

With improved winter nutrition, a 5 – 10 per cent increase in lambing percentage may be expected compared to unfed ewes. However, when ewes are forced to lamb on the open hill, little attention can be given at lambing time and losses due to exposure, starvation, mismothering and predators will contribute to lamb deaths. The next step is to provide a lambing paddock where ewes can be brought in for attention during the lambing season (see Fig. 4.3). The site for the

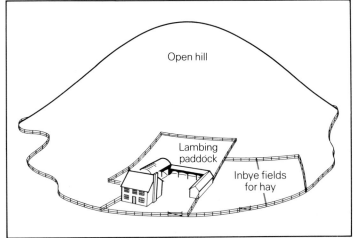

Figure 4.3 Introduction of a lambing paddock

lambs produced, bringing the overall lambing percentage to 90–95 per cent.

The introduction of improved pasture

Although some reductions in losses and improved weaning percentage can be made by winter feeding and lambing control, neither are likely to allow more than a modest increase in the stock-carrying capacity of the farm or improve the growth performance of the lambs (weaning weights). The next step in improvement is to look for increased pasture production which can allow increases in the number of ewes and the weights of lambs produced. This requires pasture improvement.

The most readily improvable areas need to be identified from the points of view of:
1. Access;
2. Requirement for drainage;
3. Soil and existing vegetation type;
4. Cost of improvement.

The areas identified may be bent-fescue areas at best or mat-grass or heather heaths at worst. The type of land selected determines which method of improvement will be used.

The area is fenced and lime and slag applied to correct the acidity and mineral deficiencies of the soil. A limited

lambing paddock should be chosen mainly for convenience of access, but it should also be a well-sheltered area, dry and free of ditches and streams.

The improvements described so far will result in improved ewe nutrition, leading to better lamb birthweights, and better control at lambing time, resulting in reduced lamb mortality. The overall result should be a 10–15 per cent improvement in

approach, by controlled grazing alone, may effect some improvement and allow a modest increase in sheep output but, where possible, far greater production is obtained from introduction of more productive grasses and clover by reseeding. In Fig. 4.4 two areas of land have been fenced and improved.

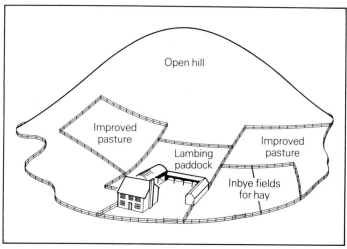

Figure 4.4 Introduction of improved pasture

A system for the utilization of improved areas (the 'two-pasture system')

The successful establishment of improved pasture must be complemented by a management system which ensures its effective utilization by grazing control. This encourages persistence of the improved sward and allows the maximum impact on animal performance. Work over the last 10 years at the Hill Farming Research Organisation at Edinburgh has centred around a 'two-pasture system' which is designed to use the improved areas most effectively in conjunction with the open hill.

The critical periods of sheep nutrition are around mating, late pregnancy and lactation. Late pregnancy demands are best met from supplementary feeding. The mating period and lactation requirements (particularly of twin-bearing ewes) can now be met from the improved pasture areas. A 'two-pasture system' uses the improved areas only at these times and allows the sward to be rested at other times while the ewe stock grazes the open hill (Fig. 4.5).

Lambing – weaning
Lambing takes place in the lambing paddock and ewes and lambs graze the improved pasture during lactation and the lamb growth period. Where there is a limited amount of improved pasture, ewes with twins graze the improved pasture while ewes with singles are turned onto the hill. Twins are

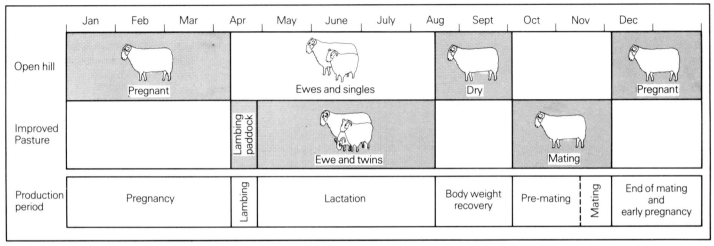

Figure 4.5   The use of improved and unimproved pasture at different times of year (The two pasture system)

always given preference for the best grass, but the eventual aim is to graze both singles and twins on improved pasture.

Summer grass production on the hill is utilized by young breeding sheep, dry ewes and cattle. In periods of grass shortage on improved land, which may occur in late summer, it may be necessary to turn ewes and lambs back to the hill for a few weeks before weaning. It is important to keep the grass short and leafy, and if the improved areas become too long the cattle may be brought in to eat off the rough grass.

An alternative practice is to allow ewes and lambs access to improved ground during the summer but to 'leave the gate open' so that they also graze the hill. They rake down into the fields in the morning and are herded back out to the hill later in the day. This allows more ewes to use the improved pasture and avoids certain mineral deficiencies (copper and cobalt) often found on improved pasture but rarely on unimproved hill land.

Where there is enough improved pasture, however, it is preferable to keep the ewes and lambs on the improved areas. Pasture treatment, the use of copper and vitamin B_{12} injection or cobalt oxide bullets can be used to overcome these mineral deficiencies.

Weaning – premating

Lambs are usually weaned in mid-August on hill farms and sold as stores (feeders). Where hay aftermaths, rape or Dutch turnips are available, some or all of the lambs may be fattened. Ewes are turned back to the hill and the improved pasture rested to allow regrowth.

Premating and mating

The ewes are returned to the improved areas in early October with the intention of improving body condition before mating. The aim is to provide a good level of nutrition throughout the mating period, but when grass becomes short in the improved areas, access can be given to the remainder of the hill. The improved areas are closed up by the end of December.

The ratio of rams : ewes is normally 1 : 40–50. The rams are withdrawn at the end of December to avoid a protracted lambing.

Winter management

In winter, the ewes must rely on the hill to provide a bare maintenance ration. This is during the period of mid-pregnancy when some decline in weight and condition is acceptable and inevitable. However, in snowy weather, ewes cannot get to the herbage and it is necessary to feed hay or sugar-beet nuts as a substitute. Lean ewes are drawn out at each handling of the ewes (dipping, injection, etc.) and kept on inbye fields to be fed hay and/or beet pulp nuts.

Prelambing

Eight weeks before lambing, blocks are laid out to ewes on the hill at the rate of one 25 kg block per 30 ewes per week. If it is possible to feed concentrates or whole maize, this may be done at selected feeding sites at a rate of 125 g/day, increasing to 250 g just before lambing. Lean ewes receive 250 g concentrates, increasing to 500 g by lambing on the inbye fields.

A week or so before lambing the ewes are brought into the lambing enclosure and receive hay or blocks plus concentrates (depending on previous level, but at least 250 g/head daily). If rams are fitted with a raddle (a coloured crayon to indicate mating), which is changed at 2-week intervals during mating, ewes may be separated into groups for lambing. They can then be brought in groups into the lambing enclosure and fed accordingly. Feeding continues on the improved area or open hill, using either concentrate feeding or blocks during early lactation, gradually reducing as grass growth begins.

Increase in production following introduction of a two-pasture system

The improvement in year-round nutrition resulting from the introduction of a two-pasture system, compared to an open-hill situation, should raise the lambing percentage above 100 per cent.

Under good conditions, many farms with a fair proportion of improved ground (20% or more) achieve lambing percentages of 110–120 per cent.

The increased herbage production affords the opportunity to increase the ewe numbers and this further elevates production. Ewe numbers are increased gradually as more land is improved and, as hill flocks are normally self-replacing, it is desirable for a gradual increase to occur by keeping more ewe-lambs. The increase in lambing percentage helps to provide this opportunity.

The way that land improvement is translated into increased sheep output has been demonstrated at the Redesdale Experimental Husbandry Farm (see Fig. 4.6). On a purple moor grass dominant sward at 310 m above sea-level, approximately 15 per cent of the area was fenced, lime and slag applied and the area stocked heavily with cattle to remove the rough vegetation and break up the surface. The area was oversown in the following summer with a mainly ryegrass and white clover mixture. The area is used according to the two-pasture system for critical periods and rested in late summer and late winter. The effect on sheep numbers and performance is shown in Fig. 4.6.

Further improvement

As the ewe numbers and performance increase, further land

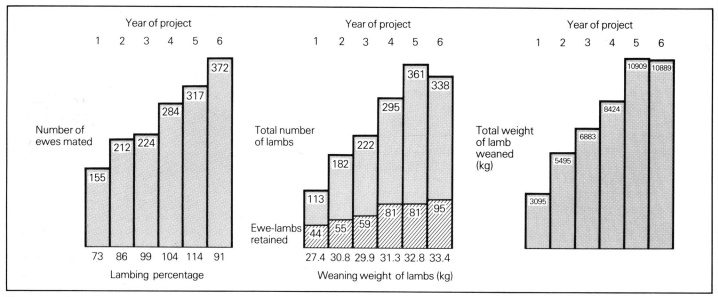

Figure 4.6 Sheep production following land improvement at Redesdale EHF (after J. R. Thompson (1978) *Blackface Journal,* 27–8)

improvement can be carried out (Fig. 4.7). The additional income from the farm is used to finance the extension of better land. The potential for further development will depend on the availability of suitable land.

Because of the increase in stock numbers, ploughing and reseeding of some land to provide fields for extra hay production should be undertaken where possible. As more good pasture becomes available, some ewes can be mated to crossing rams (like the Border Leicester or Texel) to produce crossbred ewe-lambs for sale for breeding and heavier wether lambs. Crossbred lambs command a premium over purebred lambs and the value of the product will therefore increase.

On a farm with a high potential for improvement, a hill area can be converted effectively into an upland farm. There is then the possibility of using crossbred ewes, with a higher potential performance, on the more productive pasture.

The extent of possible development will depend on the potential of an individual farm. By improving performance and stock-carrying capacity *as far as possible*, the maximum use will be made of the resources and the financial returns increased.

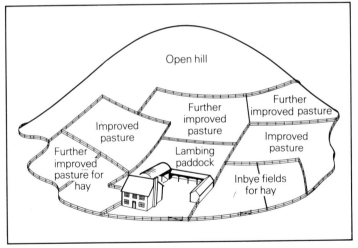

Figure 4.7 Further improvement

Feeding the ewe

5

Sheep are dependent for most of their diet on pasture, forage crops and conserved forages. For this reason, climatic differences influence the requirements for supplementation and the response to additional feed inputs. Less control can be exerted over the diet of the ewe compared to other livestock and an accurate assessment of nutrient intake is hard to make.

Hay, silage and cereal-based concentrates are usually fed to sheep as supplements to pasture and forage crops grazed *in situ*. The contribution to the diet of forages varies considerably between farms, years and seasons. The amounts of supplementary feeds required to meet the total needs for energy, protein, minerals and vitamins varies accordingly. Because we cannot measure grazing intake in practice, we must assume how much is derived from pasture and adjust supplementary feeding according to the response of the sheep, by assessing changes in weight or body condition.

The quality of feeds such as hay also varies, but this can be analysed by the advisory services and the amounts of concentrates fed can then be adjusted to match the quality of the hay.

Feed allowances for sheep should be derived from:
1. The nutrient requirements of sheep of the given weight and expected performance (e.g. lowland 70 kg ewes with twins, hill 50 kg ewes with singles).
2. The body condition of the sheep – it may be better to separate lean ewes from the majority of fit ewes in the flock and feed these at a higher level.

3. The quality of the forage fed and whether the ewes are allowed to pasture.

The annual cycle

The nutrient requirements of the ewe at any time of year cannot be determined without reference to other periods, expected changes and levels of production. There are certain times of year when adequacy of diet and body condition are critical, and other times when controlled weight loss and reduction in body condition are acceptable or even desirable.

Ewe condition at mating may affect the number of eggs shed while maintenance of condition during the first month of pregnancy is important to embryo survival. In mid-pregnancy, the ewe may be permitted to lose some body weight and condition provided that health is not affected. In late pregnancy the ewe should have sufficient body reserves to meet the moderate deficiencies in the diet which arise because intake of roughages is reduced and the contribution of pasture becomes less as winter progresses. In early lactation, weight and condition loss may occur as a normal consequence of milk production, but an increase in intake of pasture in spring means that this loss should be recovered in mid-lactation, provided that there is an *abundant* supply of grass. Weaning the lambs enables ewes to make further recovery before coming round to mating again. The cycle is summarized in Figure 5.1.

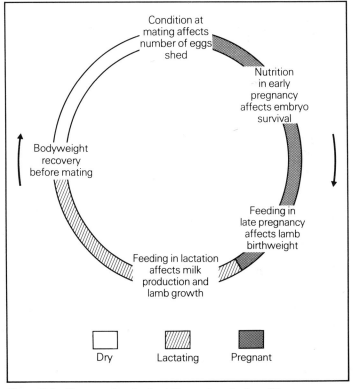

Figure 5.1 The annual cycle of nutrition and production of the ewe

Performance of sheep under extensive (e.g. hill) conditions is usually relatively low and the animals are expected to survive throughout the year on poor pasture. It is just as well, therefore, that lower condition at mating does indeed result in a lower lambing percentage, requiring less call upon body reserves in late pregnancy. Twins can be an embarrassment on the hill. The majority of the lambs will be singles so that there will be less depletion of body reserves in lactation and less need for recovery later.

Better-fed ewes will have higher body condition at mating and will produce more twins. They will have higher requirements in pregnancy and lactation. Depletion of body reserves will be greater, requiring more recovery if high performance is to be repeated in the following year. The whole system must function at a higher level.

Problems will arise when an imbalance occurs in this cycle. For example, ewes in good condition at mating will produce more twins but, if condition is allowed to decline in late pregnancy and feeding is inadequate to meet the requirements, small non-viable lambs will be born. Similarly, twin-bearing ewes may milk heavily and (particularly under hill conditions) be unable to regain body condition sufficiently to achieve twin ovulations at the next mating.

Lifetime development

The weight of the ewe increases up to about 3 years of age and some allowance must be made for weight gain, particularly in the first 18 months. When ewes are not mated before 18 months, feeding should allow a modest increase in weight in the first winter and grazing will usually provide for the necessary gain in the second summer.

Ewe-lambs which are mated have a requirement both for pregnancy, lactation and maternal weight gain. Weight gain must be made before and during early pregnancy. However, overfeeding in late pregnancy can result in large lambs within small ewe-lambs causing difficult births. Ewe-lambs which are underfed, especially if they have lambs, will be small at 18 months and less productive in the following year. Similarly, underfeeding of 18-month-old maiden ewes results in poor results in the second lamb crop.

Old ewes also present special problems, particularly when teeth loss ensues at between 4 and 6 years of age. Hill grazing and diets such as turnips are suitable for ewes which are losing teeth, and old ewes should either be culled or grazed on lowland pasture and fed diets which are more easily consumed.

Flock division and feeding practice

Ideally, ewes should be separated during pregnancy according to their expected time of lambing. This may be done if rams are 'raddled' at mating. A marking crayon is strapped by means of a harness to the breastbone of the ram and this marks the ewe's rump when she is mated. Ewes due to lamb at the start of the lambing period can then be drawn out and feeding started earlier. The different needs of ewes can thus be met more precisely.

It is also good practice to separate ewes on the basis of body condition so that leaner ewes receive preferential feeding. To do this, ewes must be 'scored' according to body condition (see Fig. 5.2). The technique relies on the principle that the loin is the last part of the body to put on fat and the first to lose it. The condition of the sheep is assessed by feeling the backbone and lumbar processes with the fingers. The sharpness of the bones, thickness of the muscle and the degree of fat cover should be assessed by feeling in the loin area, above and behind the last rib. Figure 5.2 shows the scores (1–5) that are attributed to ewes in different body conditions.

The ideal condition at mating for maximum lamb production is score $3\frac{1}{2}$. Ewes below $2\frac{1}{2}$ should be drawn out before mating for extra feeding or preferential grazing. This practice should be repeated throughout pregnancy and lean ewes drawn into a separate flock for feeding. Under extensive conditions, lower targets may be set, ewes below 2 being drawn out under these circumstances.

The lean ewes should either receive additional concentrates to permit body-weight gain or be run with the earliest lambers so that they are fed for longer. Young ewes (coming up to their first lambing) should also be fed separately from the main

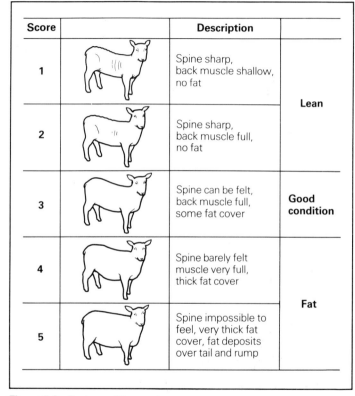

Score		Description	
1		Spine sharp, back muscle shallow, no fat	**Lean**
2		Spine sharp, back muscle full, no fat	
3		Spine can be felt, back muscle full, some fat cover	**Good condition**
4		Spine barely felt muscle very full, thick fat cover	**Fat**
5		Spine impossible to feel, very thick fat cover, fat deposits over tail and rump	

Figure 5.2   Body condition scoring

flock because they tend to compete unsuccessfully with older ewes at the trough. Old ewes which have lost teeth are likely to be in poor body condition and should be fed separately on a diet which is readily consumed, avoiding turnips.

Sufficient trough space (350–450 mm per ewe for large ewes and 300 mm per ewe for small ewes) must be provided when feeding concentrates. Hay should be fed from a suitable rack to prevent wastage and fresh water made available continuously.

Nutrition at mating

'Flushing' ewes (improving their nutrition in the 3–4 weeks prior to mating) to increase the number of lambs born has been practised for centuries. Both previous nutrition and body condition at the time of mating are important. Ewes in better body condition will produce more lambs anyway, but flushing will also increase the fertility of leaner ewes.

There are breed differences in the response to flushing. Breeds of low fertility such as hill breeds in the UK and New Zealand Romneys respond well to flushing, but more prolific breeds, including Border Leicester crosses out of hill breeds and the highly prolific Finnish Landrace and its crosses, respond less dramatically or not at all (see Fig. 5.3).

However, achieving high performance from more prolific ewes depends on the maintenance of good condition and adequate body reserves throughout the annual cycle. Good condition prior to the winter is important if ewes are to reach late pregnancy in a satisfactory state. Lowground ewes, therefore, also need to be in good condition at mating.

Management before mating is the principal factor affecting ewe weight and condition during pregnancy. In an experiment at Edinburgh with Border Leicester × Blackface ewes, two pre-mating treatments were compared. Flushing for 8 weeks resulted in heavier ewes that were in better condition at mating than ewes flushed for 3 weeks. The number of lambs produced was unaffected but the twin lambs from ewes flushed for 8 weeks were 0.7 kg heavier at birth compared to those flushed for only 3 weeks. An extra month of feeding before lambing reduced this difference to 0.4 kg but did not make up for all the effects of the earlier treatment. The results are shown in Fig. 5.4.

Good condition at mating, achieved by providing good grazing in the period before and during mating, is the target for all levels of sheep production if maximum lamb production is to be achieved.

Nutrient requirements for pregnancy

The foetal lamb makes two-thirds of its total growth during the final 6 weeks of pregnancy (see Table 5.1). Consequently, it is normal practice to increase the level of concentrate

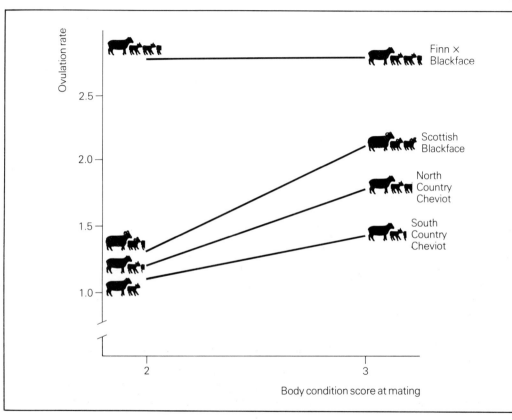

Figure 5.3 How body condition at mating affects fertiliy in different breeds (after J. M. Doney and R. G. Gunn (1973) *Hill Farming Research Organisation Sixth Report 1971–73*)

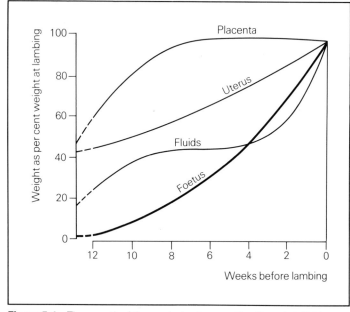

Figure 5.4 The growth of the products of conception (from J. J. Robinson, I. McDonald, C. Fraser and R. M. J. Crofts (1977) *J. agric. Sci., Camb.* **88**, 539–52)

Table 5.1 How ewe weight, condition and concentrate feeding affects birthweight of twin lambs

Weight in kg (and condition score) of ewes at mating	Birthweight of Lambs (kg)	
	12.5 kg concentrates fed over 4 weeks before lambing	25 kg concentrates fed over 8 weeks before lambing
71 (3.2)	4.7	4.7
65 (2.5)	4.0	4.3

feeding over this latter period, which will normally be in late winter when no grass growth occurs. The placenta grows during the first 90 days of pregnancy and only functional changes take place after this. Severe under-nutrition in early pregnancy can affect placental growth and may therefore affect the supply line of nutrients to the foetus later on.

Provided ewes are in good body condition at mating, they should be fed enough to maintain weight during the first 90 days of pregnancy and thereafter requirements increase with demands for foetal growth. Bigger ewes require more feed than smaller ewes and twins require more feed than singles.

The consequences of under-nutrition in late pregnancy are reduction in lamb birthweight and survival rate and the possible occurrence of 'twin lamb disease' (pregnancy toxaemia) which results in the collapse and possible death of the ewe due to the production of toxic chemicals in the blood from the rapid breakdown of fat.

Although the greatest demand is for energy at this time, adequate protein is also essential. Protein and energy interact to some extent so that their requirements cannot be separated. The effects of different energy and protein levels on lamb birthweight and ewe body-weight loss are calculated in Table 5.2.

The main danger of protein under-nutrition arises when low-quality hay is fed. When this is unavoidable, the level of concentrate feeding should be increased and a higher protein concentrate used. With good hay, protein will be adequate if a 13 per cent crude protein concentrate is used and with most silages, cereals alone will provide sufficient protein.

Table 5.2 How energy and protein intakes during pregnancy affect birthweight (kg) of twin lambs. Figures in square brackets indicate weight gain or loss by ewes from 10 weeks before to just after lambing. (Calculated from equations in J. J. Robinson & T. J. Forbes (1968) *Anim. Prod.* **10**, 297–309)

| | | Protein (g DCP/day) | | | |
		V. low (50)	Low (75)	Mod. (100)	High 125
	High (14)	4.3 [+0.7]	4.5 [+1.5]	4.7 [+2.5]	4.9 [+2.3]
	Mod. (12)	4.1 [−2.5]	4.3 [−0.3]	4.5 [+0.7]	4.7 [+0.5]
Energy (MJ ME/day)	Low (10)	4.0 [−4.4]	4.2 [−2.1]	4.4 [−1.1]	4.6 [−1.3]
	V. low (8)	3.8 [−6.2]	4.0 [−3.9]	4.2 [−2.9]	4.4 [−3.2]

Pregnant ewe-lambs

To allow maternal weight gain, pregnant ewe-lambs should receive a moderate level of concentrates (200 g/head daily), and turnips if available, throughout early pregnancy. High lamb birthweights are undesirable in ewe-lambs because they result in difficult lambings, and overfeeding in late pregnancy must be avoided. Feeding should therefore continue at the same rate (200 g/day) right up to lambing unless the weather is bad and winter pasture scarce when concentrates can be increased to 300 g/day.

Recent work in Ireland by Dr J. F. Quirke and others has shown that protein levels can affect maternal weight gain and lamb birthweight in ewe-lambs. Increased protein levels throughout pregnancy resulted in higher maternal weight gains. On a low-energy diet, high protein also increased lamb birthweight, but on a high-energy diet, high protein depressed lamb birthweight. In practice it is advisable to feed concentrates containing 13–14 per cent crude protein to ewe-lambs.

Forage quality and systems of wintering

Hay or silage usually forms the basis of winter feeding and is supplemented with concentrates and sometimes roots. Better hay and silage contain more energy and protein, are more digestible and are eaten in greater quantities. Smaller amounts of concentrate are required when good-quality roughage is provided.

Hay should be analysed before the winter starts so that the feeding can be planned. Table 5.3 shows the effect of hay quality on the concentrate requirements of lowground ewes.

When turnips and hay are fed together, the amounts should be restricted because the bulky nature of the diet may cause vaginal prolapse. Turnips are a valuable feed because they have a high energy value in the dry matter and the need for concentrates is again reduced. They may be grazed *in situ* and the ewes given access for 4–5 hours/day.

Silage can be fed to sheep, but this must be high in dry matter and digestibility, otherwise intake will be low. Introduction of drier fodder (hay) and extra concentrates should begin in the last 2 weeks of pregnancy to compensate for the reduced intake of bulky silage.

The system of wintering will also affect the amount of feed required. The merits and drawbacks of three systems of winter feeding are illustrated in Fig. 5.5. Ewes with access to pasture in winter will derive some benefit from grazing and will require less forage feed (hay or silage). When ewes are housed or on very limited pasture, all the feed requirements must be met from supplementary forage and concentrates. The same applies when there is snow cover or in stormy weather.

Table 5.3 Effect of hay quality on concentrate requirements

Hay quality	Energy (MJ/kg DM)	Protein (DCP g/kg DM)	Concentrates required by lowland ewe (g per head daily)
Poor	7.5	20	225 ——— 900 over last 8 weeks
Average	8.5	40	225 ——— 675 over last 6 weeks
Good	9.5	60	225 ——— 450 over last 4 weeks

Nutrient requirements for lactation

Current nutrition is the primary factor influencing milk production of ewes, although maximum production is also dependent on adequate nutrition in both pregnancy and lactation. Feeding during the first 4 weeks of lactation is particularly critical and affects the subsequent lactation performance of the ewe and hence growth of the lamb.

The demand for energy in lactation is very high and requires that substantial levels of good-quality food are provided. Spring grass is satisfactory but, when this is not yet

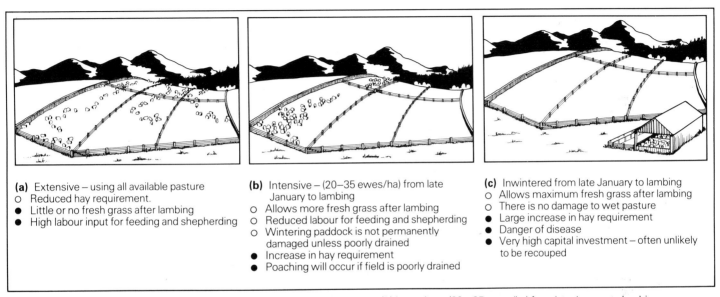

(a) Extensive – using all available pasture
○ Reduced hay requirement.
● Little or no fresh grass after lambing
● High labour input for feeding and shepherding

(b) Intensive – (20–35 ewes/ha) from late January to lambing
○ Allows more fresh grass after lambing
○ Reduced labour for feeding and shepherding
○ Wintering paddock is not permanently damaged unless poorly drained
● Increase in hay requirement
● Poaching will occur if field is poorly drained

(c) Inwintered from late January to lambing
○ Allows maximum fresh grass after lambing
○ There is no damage to wet pasture
● Large increase in hay requirement
● Danger of disease
● Very high capital investment – often unlikely to be recouped

Figure 5.5 Systems of wintering ewes (*a*) Extensive – using all available pasture (*b*) Intensive – (20 – 35 ewes/ha) from late January to lambing (*c*) Inwintered from late January to lambing

available, ewes will require concentrate supplementation at an even higher rate than in late pregnancy. Lactation usually makes some demand on body reserves, and recent evidence from the Rowett Institute indicates that the *energy* loss may be even greater than is suggested by the loss in body weight (because fat is used and replaced with water). Ewes must therefore be in fairly good body condition at lambing, especially if they are to rear two lambs, as well as being fed a

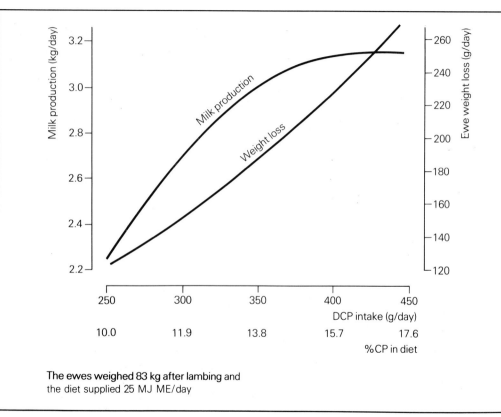

Figure 5.6 The effects of protein intake on milk production and live-weight loss in ewes receiving a high energy diet (from J. J. Robinson, C. Fraser and J. J. Gill (1973). Paper presented at the British Society of Animal Production Meeting, Harrogate, March 1973)

The ewes weighed 83 kg after lambing and the diet supplied 25 MJ ME/day

high level of energy in early lactation.

Again there is an interaction between protein and energy in the response of ewes in lactation. Figure 5.6 shows how ewes on a low-protein diet produce less milk but also lose less weight than ewes on a high-protein diet but the same energy level. A high-protein diet encourages milk production at the expense of body reserves. The diet for lactating ewes should therefore contain 16–18 per cent crude protein.

The value of by-pass proteins

As a ruminant the ewe depends on micro-organisms in the rumen for the digestion of food. The microbes break down cellulose, starch and sugars to volatile fatty acids and protein to simple nitrogen compounds. They also resynthesize protein from simple nitrogen sources. To perform these functions, the microbes themselves need a balanced supply of energy and nitrogen (which can be supplied) either from protein or urea).

Recently, it has been found that, even at high energy levels, milk production may be increased by supplying more protein than can be provided by the rumen microbes. To do this proteins can be fed which are not broken down in the rumen but pass through for digestion in the duodenum. Fishmeal is the ideal 'by-pass protein' (about 70 per cent is un-degraded in the rumen). Soya bean and linseed meal are better than groundnut meal for this purpose, but urea obviously provides no protein except that synthesized by the microbes.

For maximum milk production, we must satisfy both the requirements of the microbes and the ewe itself. A high energy level, some degradable protein and additional non-degradable protein are required. A cereal diet supplemented with fishmeal (plus minerals and vitamins) is the ideal concentrate for early lactation. Feeding should be gradually reduced as grass grows as sudden withdrawal of fishmeal can cause hypocalcaemia.

A feeding plan for lowground ewes

The principles discussed so far can be combined into a plan for feeding the lowground ewe aimed at producing and sustaining two viable lambs.

1. In late summer (August) 50 kg N/ha is applied to provide extra grass for flushing. Pasture is rested after weaning, but ewes are dosed for worms and returned to good grass fields in early September to improve body condition before mating. Any lean ewes are given preference for the best grass fields or young grass (undersown stubbles). Any ewe-lambs for mating will benefit from similar treatment.

2. Body condition is checked again at mating. Lean ewes are added to the special group. If grass is scarce, the lean ewes receive 225 g concentrates from mating onwards.

3. From mid-winter (January) onwards, ewes are confined to a small area of permanent grass or next year's hayfield to allow the majority of the grass to freshen up for spring

grazing. Good-quality hay is fed. Lean ewes are drawn out and run in a separate field; they receive good-quality hay plus 225 g concentrates throughout.

4. Ewes are given access to turnips (if available) from 6 to 8 weeks before lambing for 4–5 hours/day. Concentrate feeding depends on hay quality. With average hay it begins 6 weeks before lambing at 225 g/head daily. The level is increased to 450 g 3 weeks before lambing and 675 g 10 days before lambing. Lean ewes are fed 225 g/head more than the rest of the flock until the last 10 days when all ewes are fed 675 g daily. If ewes can be separated by mating date (using crayon markers on the rams) feed levels can be better related to lambing date and lean ewes may be run with the earliest lambers.

5. By lambing, lean ewes are in reasonable condition and can return to the main flock for lambing. All ewes receive 675 g concentrates with good hay. Turnips can be laid down in the lambing field.

6. Ewes are turned out onto a restricted area to allow spring grazing to freshen up, but gradual stocking up of the summer grazing fields begins with the oldest lambs. Before grass growth takes off, 675 g concentrates per head daily is fed with *ad lib.* turnips. Ewes normally consume little or no hay after lambing. As grass availability increases the concentrates are reduced to 450 g, then to 225 g and finally to nil. At this time grass will be growing rapidly. Turnips continue to be fed until the supply is used up.

Winter feeding of hill ewes

Because of the lower level of production, feed inputs to hill ewes are limited by economic constraints but requirements are lower. Hill sheep can derive most of their needs from natural grazings, but these must be augmented by feeding conserved forage (hay) in bad weather and snow. Better grass (improved hill grazing or adjacent grass fields) can be used at mating and in lactation, and forage crops can also be grown for feeding at these times.

A low level of concentrate feeding is justified in late pregnancy, but the main problem is the practicality of feeding under extensive conditions. It is undesirable to disturb the natural grazing behaviour of the flock as they must derive most of their needs from pasture. Feeding must be carried out on the normal sheep-grazing areas and feed stored in weatherproof containers near to the site.

Feeding may be done on alternate days to reduce the tendency for sheep to loiter at the feeding site. Hard pellets or whole grains, such as maize, are feeds which can be scattered on the ground so that troughs are not required. An alternative is to use feed blocks. The latter are complete feeds in a hard block with appetite inhibitors included to limit intake. They can be distributed weekly to convenient sites on the hill.

Under extensive conditions, 85 per cent or so of requirements will normally be derived from pasture or body reserves. The aim of supplementing hill ewes is to prevent the depletion of body reserves from going too far. This will have serious repercussions at later stages in the annual cycle. It must be remembered that low levels of concentrates or feed blocks will only meet the 15 per cent of requirements for which they are intended. This emphasizes the point that supplementary roughages must be provided when pasture is covered by snow or in bad weather. The economic level of feeding is determined by the expected performance, but as the latter is dependent to some extent on the former, a decision must be taken at what level the system is required to operate.

Feed blocks for hill sheep

Feed blocks have proved an excellent method of providing supplementary feed on inaccessible hill ground with a minimum of disturbance to natural grazing patterns. However, farmers should be aware of the limitations if good results are to be achieved.

Firstly sufficient blocks should be provided to achieve intakes of about 115 g/head daily. The manufacturers' recommendations are generally one block per 25–30 ewes/week. The actual amounts eaten vary widely and should be checked on the individual farm. Sheep will eat more if extra blocks are laid down.

Secondly, the blocks should be carefully sited at strategic points on the hill, particularly for ewes after lambing, where they can be reached without crossing burns or ditches.

At the recommended levels blocks are only meeting 10–15 per cent of the energy requirements of the ewe at critical times such as late pregnancy and early lactation. The remainder must come from hill pasture and body reserves. In bad weather, particularly when there is snow cover, additional hay must be provided to compensate for the lack of natural forage. Thin ewes are probably those which are not eating the block and these should be brought inbye and fed conventional concentrates and hay.

The choice of block is difficult with the wide range now available. It may be necessary to experiment to find the type which works best on a given farm, but the choice is best governed by the energy value of the block in relation to its cost. Many blocks contain a high level of urea which can be converted to protein in the rumen. It is claimed that this stimulates roughage intake but, while this may be the case with a predominantly heather diet, it is not true of most hill grasses which generally contain sufficient protein for the rumen microbes in any case. Hence energy is the most important consideration.

Feed blocks cost about twice as much per unit of energy compared to barley or maize and where it is possible to feed cereals (supplemented with protein and minerals) this should be done in preference. The extra cost of blocks is justified by the fact that they reduce the problems of feeding under difficult conditions on the open hill (Fig. 5.7).

Figure 5.7   Feed blocks for hill sheep in Wales

A feeding plan for hill ewes

The practices best suited to an individual hill farm will depend on the particular environment and expected performance. The following plan assumes a target of 100 per cent lambing is attainable.

Ewes are handled at every gathering (dipping, mating, vaccination, etc.) and lean ewes drawn out and kept on fields near the farmhouse.

1. If the hill has enclosed areas which have been improved or sufficient inbye ground, ewes are brought in for flushing. If mating has to take place on the open hill, ewes are handled to check body condition and any lean ewes kept in fields.
2. Ewes on the hill require supplementary feeding with hay (up to 675 g/head) when there is snow cover. Block or concentrate feeding commences about 8 weeks before lambing (30 ewes/block/week or 225 g concentrate per head daily) and continues until after lambing. Lean ewes are kept in fields and receive hay (675 g) and concentrates (375 g) per head daily. The concentrates are increased gradually to 450 g 2 weeks before lambing.
3. On the open hill, blocks or concentrates and hay are fed before and during lambing time. If an enclosed area or field can be used, ewes are brought in before lambing and fed blocks or concentrates (450 g) and hay. Ewes are wormed at this time and an anthelmintic block may be used if conventional drenching is impractical.
4. Feed blocks continue to be available until grass growth starts (May). If twins can be kept on improved ground or inbye fields, performance will be better and concentrates (450 g reducing to 225 g) may be trough fed.

Feeding hill ewe-lambs

Ewes should be taught to eat blocks and/or concentrates in their first year (preferably in early autumn). Severe nutritional stress as a young sheep reduces the subsequent performance as a ewe so that good wintering is essential. Up to 675 g of hay should be given to young breeding stock when grass is scarce or in snow, and a low level of concentrates (110 g) should be fed under these conditions.

Management at lambing

6

The lambing field

Hill farmers with no fields or enclosed areas are forced to lamb on the open hill. Lambs born in this environment are vulnerable to bad weather, predators and mismothering; shepherding is virtually impossible. If lambing takes place within a confined area, attention can be given to weak or lost lambs and ewes in difficulty can be assisted. Fencing of a dry but unimproved area of land, using cheap electric fencing, provides a suitable lambing area and keeps pressure off fields near the steading and good grass until after lambing. A confined area is more favourable to the spread of infectious disease among ewes and lambs, but the risks of death from exposure and starvation are far greater than those from infection. Care in disease prevention and hygiene precautions are essential in the lambing area.

Upland and lowland farms generally provide a lambing field which is sheltered and accessible to the shepherd. Many farms use the same field each year and, although this incurs the danger of carrying over disease from one year to the next, the advantages of the most convenient and sheltered field usually outweigh this risk.

The use of a building to house the lambing ewes, either continuously or at night, is common practice on lowland farms. Disease problems are greatest in the housed environment and a build-up of bacterial infection (especially *Escherichia coli* which causes scour and 'watery mouth' in young lambs) often occurs.

If different fields are used for lambing each year, the problems of disease are reduced. A temporary shelter which can be moved each year is often better than a permanent lambing shed. If disease problems start to build up as lambing progresses, the flock should be moved onto a clean field. When planning the lambing operation, a second area should be identified to use in such circumstances.

Lambing groups

A smaller number of ewes in the lambing field allows closer supervision and reduces the danger of disease build-up. If ewes are brought into the lambing field only when they are about to lamb, the numbers in the field at any one time are kept to the minimum. At mating, a wax crayon attached by a leather harness to the breastbone of the ram makes a mark on the ewe that indicates when mating has taken place. The colour crayon is changed every 7–10 days. Adding 145 days to the date of mating allows the approximate date of lambing to be determined. A more permanent mark can be made with marking fluid to indicate when the ewe is expected to lamb.

In this way, ewes can be separated according to lambing date to allow more accurate feeding and so that they can be brought into the lambing field in weekly groups just before they are due to lamb.

Lambing pens and shelter

A sheltered field with a covered area or shed for the night is ideal (see Fig. 6.1). Additional shelter can be provided by placing straw bales around the field or tying sacking along the fences. There must be a source of water and preferably also electricity and light in the lambing area.

Ewes with twins are best kept in lambing pens for the first night. One lambing pen, 1.5×1.5 m, is needed for every 8–10 ewes in the flock. It also helps to have one or two 'hospital' pens in a warm place and preferably with a heater or infra-red lamp.

Ideally, lambing ewes should be checked regularly throughout the day and night. Ewes may require assistance, lamb navels need dressing (with tincture of iodine) and lambs may require feeding or bringing into a 'hospital' pen. Mismothering may occur if several ewes lamb at the same time and are not put in separate pens.

Lambing pens should be cleaned out and deep-bedded with straw after each ewe. Particular attention should be paid when ewes abort or lambs scour in the pen. Afterbirths and dead lambs should be removed, using protective gloves, and collected in a strong plastic bag for burning. The pens should be used in strict rotation so that a few pens are not used repeatedly.

After lambing, when the lambs are strong enough and weather conditions are favourable, the ewes and young lambs

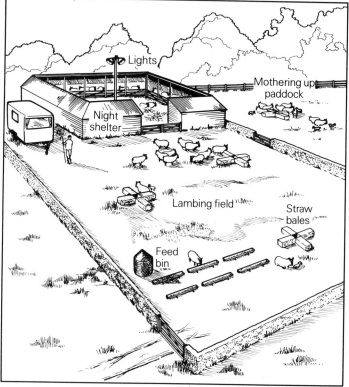

Figure 6.1   Lambing arrangements

can be turned out into a small field or paddock where they can continue to be observed and the lambs mothered up for the first few days. Additional shelter is again provided with straw bales or sacking.

Thereafter, the ewes can be moved to grazing fields by age.

Lamb losses

Average lamb losses around birth are about 8 per cent in lowground and upland flocks and considerably more on hill farms. Under good conditions losses may be below 5 per cent, but in bad years they may exceed 20 per cent. Lamb deaths deplete the total number of lambs for sale and therefore directly reduce output from the sheep flock. Improved management reduces mortality and boosts profits.

Causes of death

The relative importance of different causes of death is shown in Fig. 6.2. About a third of the dead lambs are aborted prematurely or die during birth, and the causes are likely to be physical stress, inadequate feeding of the ewes or disease. After birth, the largest proportion of deaths result from starvation and exposure: small lambs that receive inadequate milk from the ewe succumb to the effects of wet, cold and wind. Infectious diseases account for only about 15 per cent of

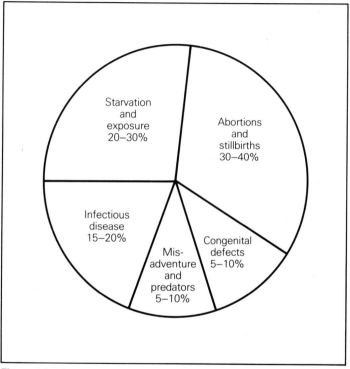

Figure 6.2   Relative importance of different causes of lamb deaths (based on surveys by MLC, the West of Scotland Agricultural College and the East of Scotland College of Agriculture)

all deaths and management factors are of far greater importance.

Abortions and still-births

'Still-born' means that a lamb is born dead and nothing can be done to save it. The lamb has died inside the ewe or during the birth process. 'Abortion' describes premature birth before the lamb is due and at a time when it is unable to survive outside the ewe. Because abortion may be caused by a number of different infectious diseases (enzootic abortion, toxoplasmosis, tick-borne fever, salmonellosis, vibriosis and others) it is important that a veterinary diagnosis is sought when abortions occur. The vet will require the aborted lamb and the afterbirth for his investigation. Both still-births and early abortions can result from physical stress or mishandling. Death may also occur as the result of a difficult birth because the lamb is wrongly positioned or too large. The latter is usually just chance, but may result from gross overfeeding in late pregnancy, particularly in young sheep. Some breeds are more prone to difficult lambing than others.

Prevention of abortion

Enzootic abortion can be prevented by vaccination and once it has been diagnosed in a flock all ewes should be injected with the vaccine. Thereafter it is necessary only to vaccinate replacements each year (usually in September/October – well before mating). Ewes which have aborted may be kept in the flock but should be kept clear of maiden ewes until discharges have ceased.

An effective control method for toxoplasmosis is not yet available. It is usually recommended that maiden ewes are mixed with the flock early so that they are exposed to the disease and can develop resistance before becoming pregnant. Another cause of abortion, vibriosis, is a less common disease and occurs only spasmodically. To prevent its appearance and limit its effects sheep should be kept on clean pasture and not overcrowded during late pregnancy.

Tick-borne fever can cause abortion if ewes encounter ticks for the first time during pregnancy. Resistance is again built up if they are exposed to ticks in early life. On affected farms, replacement ewe-lambs should be run on infested ground in autumn to allow them to develop immunity before their first pregnancy begins.

Salmonellosis is more likely when conditions are overcrowded. It is less likely to occur if ewes are kept on a clean field before lambing and contact with other livestock (particularly bought-in calves) is avoided. There is less chance of the disease spreading in small groups.

In general, prevention of infectious abortion is helped by avoiding high stocking rates of large numbers of ewes in winter, mixing of ewe-lambs with older ewes before mating and vaccination of replacements before mating with enzootic abortion vaccine (where this is diagnosed as a problem in the flock). It is essential that aborted ewes are isolated and aborted foetuses and afterbirths burnt.

Mishandling

Feet should be checked and foot ailments corrected well before lambing so that unnecessary handling is avoided when the ewes are heavily in lamb. If handling is essential at this time, ewes should not be crushed together and the pens should be designed with this in mind.

Spring dipping against tick is liable to precipitate abortions or premature births. There is evidence that early exposure to tick-borne diseases may be desirable in minimizing their later effects. Many farmers in tick areas have found that dipping at this time is unnecessary. Dipping against lice can be carried out much earlier to avoid manhandling ewes in late pregnancy.

Heavily pregnant ewes should be kept off fields with treacherous bumps or ditches and well away from houses where pet dogs are liable to worry them.

Difficult births

Difficulty at lambing can result from oversized lambs or malpresentation (wrong position). Overfeeding in late pregnancy can cause oversized lambs, particularly in ewe-lambs and young ewes. Ewe-lambs that are mated need to be fed for longer than the rest of the flock and should not receive more than 300 g of concentrates in the last 2 weeks.

Extra feeding of lean ewes and more moderate levels to fit sheep will avoid the problem of some ewes getting very fat.

When to assist

If the lamb is wrongly positioned or too big to pass through the birth passage, the ewe requires assistance. This usually occurs in less than 10 per cent of births and unnecessary assistance should not be given. Normal parturition takes about $1\frac{1}{2}$ hours from the start of straining. If the ewe is seen to strain for a long time without result, fluids or part of the lamb appear without further progress or the ewe gives up after straining for some time, it is wise to investigate the reason. All manipulations should then be gentle and under hygenic conditions because the ewe is very susceptible to injury and infection at this time. Nails must be cut, hands must be thoroughly clean and a lubricating cream used. After a difficult lambing an antibiotic injection is given or a pessary is inserted into the birth passage.

If there is any doubt about a particular problem, the vet should be called. Inexperienced personnel can be trained by an experienced shepherd or attend one of the relevant courses (organized by the Agricultural Training Board in the UK).

Assisting birth

The normal position of the lamb is shown in Fig. 6.3. If the presentation is normal but the ewe is still unsuccessful within a further half-hour, the lamb is delivered by gently straightening each leg and then easing the head out by gently pulling the exposed legs down towards the udder. Gentle rotation of the lamb may help to release it.

If one or more of the legs is backwards (Fig. 6.4), the ewe is turned on her side with the offending limb upwards and the leg straightened, cushioning the hoof in the fingers (to prevent damage to the womb). It is sometimes necessary to push the lamb gently back inside the ewe before the leg can be pulled straight. If the head is turned round (Fig. 6.5) the lamb is pushed gently back until it can be straightened. In particularly difficult cases, a lambing cord can be used (Fig. 6.6) pulling gently on the cord while tugging alternately on the two legs. But this device requires experience and skill in use and should not be tried without expert instruction.

Breech presentations (Fig. 6.7) cause the worst problems. A vet should be called if the shepherd does not feel competent or if a large lamb is involved. Plenty of lubricant should be used and the lamb eased back inside the ewe before attempting to turn it. If this fails, it must be drawn out backwards.

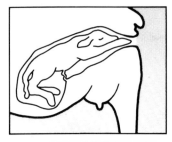

Figure 6.3   Normal presentation

Figure 6.4   Forward presentation with one or both front legs back

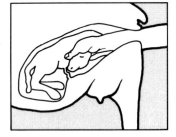

Figure 6.5   Forward presentation with head back

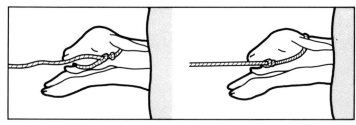

Figure 6.6   Use of a lambing cord

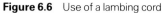

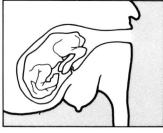

Figure 6.7   Breech presentation. The hind legs must be straightened before the lamb can be drawn

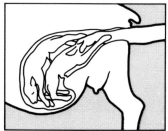

Figure 6.8   Where there are twins it is essential to identify limbs before attempting to draw the lamb

Sometimes one lamb of a pair of twins comes backwards (Fig. 6.8). The limbs belonging to each lamb must be identified before attempting the delivery. The lamb must always be pulled downwards (toward the udder).

Small and weak lambs

After birth, most deaths are caused by starvation and exposure. Small lambs are most vulnerable because:
1. they have a larger surface area relative to their weight, resulting in greater heat loss;
2. they have fewer fat reserves to utilize for heat production;
3. twins and triplets get less colostrum from the ewe due to competition and the limited supply.

The relationship between birthweight and mortality is shown in Fig. 6.9. Lambs below 2 kg at birth have only a 10 per cent chance of survival. The survival rate improves as the average birthweight is approached (4–5 kg). Very large lambs cause problems of difficult lambings.

The importance of colostrum (early milk)

A lamb must get an adequate amount of colostrum from the ewe as soon as possible after birth. Colostrum has three functions:

1. It provides a source of energy to nourish the new-born lamb.
2. It provides antibody proteins which are absorbed through the lamb's gut to give resistance to disease.

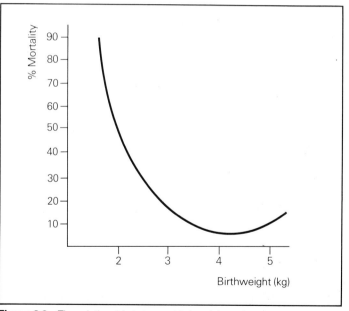

Figure 6.9   The relationship between birthweight and mortality

3. It acts as a laxative, helping to unblock the gut.

An adequate supply is particularly important for a small, weak lamb and if, for any reason, the lamb is unable to obtain sufficient from the ewe, it may be given a supplementary supply of colostrum by stomach tube (Fig. 6.10).

Weak lambs should be given 50 ml of colostrum by stomach tube. Colostrum may be milked out of another ewe and used fresh or stored deep frozen ($-17\,^\circ$C) to be warmed up when required. Cow colostrum can be used but, although this provides energy, it does not provide sheep antibody proteins. A patent device is available, or the type shown in Fig. 6.10 can be made up from a disposable plastic syringe and 250 mm rubber catheter tube (blunt ended to avoid choking).

Having filled the syringe, the lamb is supported on the person's knee in a position similar to the natural suckling stance (with the head up). The tube is passed gently down the throat – if it goes in easily there is little danger of choking, if it does not, it should be pulled out and a second attempt made. The colostrum is passed slowly down the lamb's throat by pressing gently on the plunger. The tube is removed and the lamb put in a warm place or under a heat lamp. The tube must be washed after each feed. Normally, lambs will be up on their feet within a couple of hours and can be returned to the ewe. If the lamb has to be fostered, it can be given tube feeds every 4 hours until a suitable ewe is available. Lambs fed with the tube suckle more readily when fostered because they are not

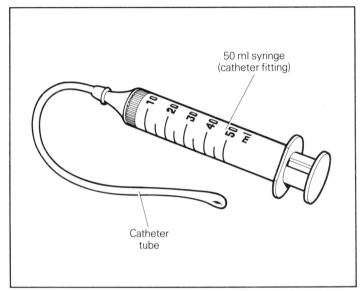

Figure 6.10   A simple stomach tube

psychologically attached to the bottle. The tube is not intended for nursing orphan lambs which can suckle quite adequately from a bottle.

Fostering

If a ewe dies or has triplets or quads, it may be necessary to foster a lamb onto another ewe. The adopting ewe needs plenty of milk and may be one that has lost its lamb or produced a single. Smaller lambs should be left with their mother and the strongest member of a set removed for fostering.

A fostering pen (Fig. 6.11), where the ewe is held by the neck, is recommended as the best routine method of fostering:
1. The lamb should have received adequate colostrum.
2. The ewe is placed in the pen and its head secured.
3. The lamb is placed in the pen with an infra-red lamp above it (unless the lamb is particularly strong and lively).
4. The lamb should be *seen* to suckle and the belly should be full.
5. The ewe and lamb are left in the pen for 48 hours before releasing the ewe's head and removing one creep bar. The ewe and lamb remain in the pen for a further 24 hours (individual ewes respond differently and time will have to be varied).

An old technique is that of skinning the dead lamb from the adopting ewe and tying the skin onto the orphan to be fostered. Fostering pens have proved more reliable and effective.

Disease prevention

Disease is not a major cause of lamb deaths on most farms. Preventive medicines and sensible hygiene precautions usually prevent serious outbreaks of disease. The main diseases that cause lamb deaths are enteritis (*E. coli*), navel infections, dysentery and pneumonia (pulpy kidney and tetanus can cause further deaths later).

Clostridial diseases (pulpy kidney, tetanus, braxy, black disease and lamb dysentery)

Breeding ewe lambs and maiden ewes are vaccinated against clostridial diseases in September or October and all ewes receive a booster injection at least 2 weeks before lambing. A 'seven-in-one' vaccine is produced and this may also incorporate *Pasteurella*-pneumonia vaccine. The immunity conferred upon the ewe is passed to the lamb via the antibody proteins in colostrum and gives the lamb immunity for the first 10 weeks of life.

Swayback (neonatal ataxia)

Swayback is the result of copper deficiency in the diet of the ewe and cannot be corrected in the lamb after birth. Where it is known to be a problem, and confirmed by soil and/or blood analysis for copper, the ewes may be injected or drenched with

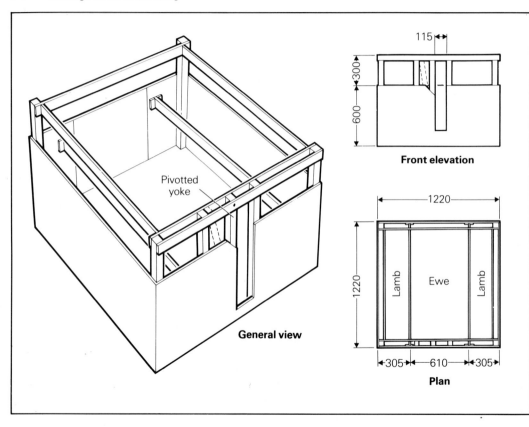

115

300

600

Front elevation

1220

1220

Lamb

Ewe

Lamb

305   610   305

Plan

Pivotted yoke

General view

Figure 6.11   Design for a lamb fostering pen

a copper preparation in mid-pregnancy. The problem is particularly acute in mild winters when swayback warnings are given by local Government agriculture departments. As copper is toxic to sheep at relatively low levels, feeding of copper containing minerals or treatment of pasture with pig slurry must be avoided, particularly when ewes are also injected with copper.

Joint-ill (tetanus)

Joint-ill is prevented by dressing lambs' navels at birth with iodine or antibiotic spray. Lambs can be tailed and ram-lambs castrated at this time, using rubber rings.

Scour and watery mouth (coliform enteritis)

Good hygiene precautions are essential at lambing time to prevent the build-up of bacterial infection leading to scouring and watery mouth. An outbreak of infection can be controlled by oral dosing with antibiotics or other proprietary preparations or by moving the flock to a clean field or shed for lambing. Choosing a clean lambing field each year is desirable, but shelter is often the first consideration and the choice may be limited. Lambing pens and sheds must be thoroughly disinfected. The pens should be cleaned out or deep-bedded between ewes, and diseased afterbirths and dead lambs collected in a plastic bag (using rubber gloves) to be burnt away from the lambing area.

Rearing orphan lambs

Inevitably, some lambs are orphaned and no ewes are available for fostering. In these cases, it is necessary to rear the lambs artificially. This is rarely an economic proposition, but it can be worth while provided there is no cost to the labour input. Fostering is preferable to bottle feeding! When artificial rearing is required, the following rules apply:

1. The lamb must receive adequate colostrum.
2. It is placed in a warm, draught-free pen on its own or with others of the same age. New-born lambs must not be put in with older lambs.
3. The lamb is left to get hungry for up to 12 hours before training it to drink from a teat of a bottle or bucket.
4. The minimum of human interference is desirable. The lamb is placed on the teat and left alone to suckle. A hungry lamb soon learns to suck.
5. Feed fresh milk substitute (a proprietary ewes' milk subsitute) made up to maker's instructions – usually one part of dried milk to five parts water – is fed either *ad lib.* or in two or three feeds.
6. Creep feed (rolled barley, flaked maize and protein crumbs) is made available from 7 to 10 days of age and the lamb is

weaned abruptly at 30 days old. Concentrates are fed right up to slaughter, changing to whole barley with a protein supplement at about 6 weeks.

Not all these will be possible under extensive (e.g. open-hill) conditions, but the aim should be to get as near to the ideal system as possible within the limitations of the given farm.

Summary of good lambing management

The main points detailed in this chapter are summarized in Table 6.1. After lambing, ewes and lambs should be checked daily to ensure all lambs are well mothered. The suggested practices are the ideal which will result in minimum losses.

Table 6.1   Summary of lambing management

Before lambing
   Check feet at least 8 weeks prelambing.
   (Give copper treatment where there is risk of swayback.)
   Vaccinate against clostridial diseases at least 2 weeks before lambing.
   Minimize handling.

At lambing
   Separate early and late lambers to limit numbers in the lambing field or
      shed.
   House at night if possible.
   Provide one lambing pen for every eight ewes lambing.
   Check ewes' milk supply after lambing.
   Foster orphan and triplet lambs using a lambing crate.
   Ensure lambs obtain colostrum (by tube if necessary).

After lambing
   Ensure all lambs are well mothered.

Growth and finishing

7

Profitable lamb production is the result of producing an acceptable product at the right time in the right place. A knowledge of the market combined with an understanding of the factors affecting growth and fattening will enable the producer to plan his production system to meet the requirements of the market and obtain maximum economic returns.

Market requirements differ both in the size of carcase and the level of fatness that is acceptable. This is illustrated by the carcase weights required in different European markets, shown in Fig. 7.1. Less fat is preferred on the Continent compared to that tolerated by the UK home market. Within the UK, heavier lambs are preferred in the north than in the south and the market for light lambs is greatest in the industrial centres.

Requirements and prices also vary with the time of year; highest prices per kilogram are paid for lightweight New Season lamb at Easter and the price gradually falls thereafter until midsummer when the numbers and weights of lambs increase and the price per kilogram drops markedly. With the plentiful supply of lambs in the autumn and winter, prices remain low before gradually increasing again in the New Year. In recent years, because of the demand for lambs for export to the Continent where prices are higher, a premium has been paid for light carcases in the autumn and winter when the majority of lambs are too heavy and fat for this market.

The value of a lamb is determined by the price per kilogram

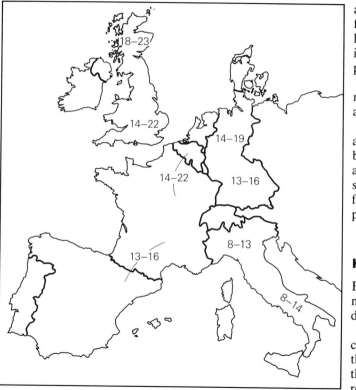

Figure 7.1   European Market Requirements – Lamb – kg. Carcase weight

and the weight of the carcase. When a large premium is paid for light carcases, the same price is obtained for a light as for a heavy carcase but, if this premium is small, the heavy carcase is worth more even if a slightly lower price per kilogram is paid.

Planning depends on forecasting the price; the forecast must take account of the normal seasonal price trend with allowances for inflation and other long-term trends.

The marketing date is determined by the time of lambing and the growth rate of the lambs, the latter being affected by breed, system of production and level of feeding. The appropriate breeds and production methods must be chosen so that the lambs will be produced with the desired level of fatness at the weight that will obtain the highest expected price per lamb.

How lambs grow

Figure 7.2 shows how growth accelerates in the first few months, then slows down towards puberty and continues to decline progressively as maturity is reached.

The total weight of the lamb is made up by the weight of the carcase and that of the non-carcase components which include the head, feet, fleece, internal organs, gut and gut contents. As the lamb increases in weight and age, the weight of the carcase relative to the non-carcase components increases.

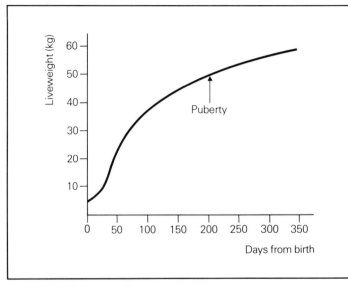

Figure 7.2 Typical growth curve of Suffolk cross lamb

which type of lamb will produce the desired carcase on different finishing systems.

The relative proportions of muscle (lean) and bone vary between breeds, but are little affected by sex or level of feeding. However, the relative rate of fat deposition is markedly affected by breed, sex and feeding. Furthermore, fat occurs both on the outside of the carcase (subcutaneous fat) and between the muscles (intramuscular fat) and other parts of the body (internal fat); the relative proportion of subcutaneous fat also varies with breed.

The shape or 'conformation' of the carcase changes as the lamb grows. The body becomes larger in relation to the head and limbs and becomes thicker and deeper. The loin area becomes better developed relative to the neck and chest and the leg joints become rounder and fatter. There are important differences in conformation between breeds, but it has been noted that these are not necessarily indicative of differences in lean content or the relative proportion of high-priced joints in the carcase. Variation in conformation is largely a reflection of differences in fat cover (subcutaneous fat) and it therefore bears a close relation to fatness. A leg of lamb can be long and narrow or short and thick and, provided the fat cover is similar, there will be little difference in lean content. Although considerable importance is attached to conformation by the meat trade, the fatness of the carcase gives a better guide to the lean meat content than conformation.

The carcase itself is made up of bone, muscle and fat. As the lamb grows, the proportion of muscle and bone decreases as the proportion of fat increases. The ratio of muscle to bone also increases (see Fig. 7.3). The pattern of growth must be known for each breed, sex and feeding regime to determine

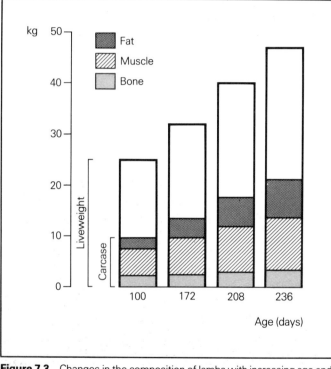

Figure 7.3 Changes in the composition of lambs with increasing age and liveweight (from T. H. McClelland, B. Bonaiti and StC. S. Taylor (1976). *Anim. Prod.*, **23,** 281)

Carcase classification

The required weight and fatness of a carcase depends on the market outlet. These requirements can change – indeed are currently changing in the UK as greater emphasis is being placed on leaner lamb. It is therefore helpful to classify lamb

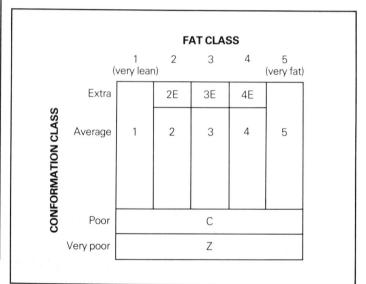

Figure 7.4 The MLC Sheep Carcase Classification Grid

carcases in a way that can be used to describe the product to potential buyers and guide farmers as to what to produce. The MLC Sheep Carcase Classification System was developed for this purpose.

The system is shown in Fig. 7.4 and examples of carcases in various classes are illustrated in Figs 7.5 and 7.6. The system is operated by a national team of trained assessors or 'graders' who score the carcases subjectively on the basis of fatness and conformation. Although there are 10 different classes, 95 per cent of carcases produced in the UK fall into fat classes 2, 3 and 4 and 75 per cent are of average conformation with the

further 20 per cent being of extra conformation. The remaining classes are generally unacceptable and only classes 2, 3, 4, and 2E, 3E and 4E are of commercial importance.

The relationship between fat class and dissectable fat, lean, bone and waste is shown in Table 7.1, which shows that there is a difference of about 6 per cent in total fat and 4 per cent in total lean between the classes. The requirement for the UK home market is for carcases in classes 3 and 3E with classes 3E and 4 acceptable in the north, whereas fat class 2 is appropriate for export carcases. Even in the UK, there is a trend towards the lower fat classes.

Fat class: 1   2   3   4   5

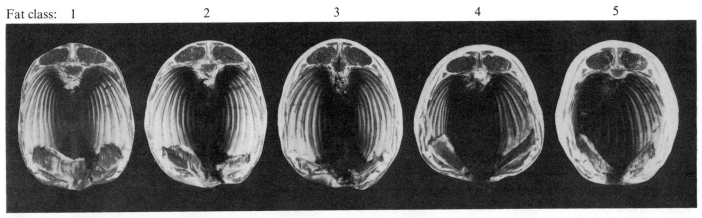

Figure 7.5   MLC Sheep Carcase Classification System – fatness: cross sections through the carcase

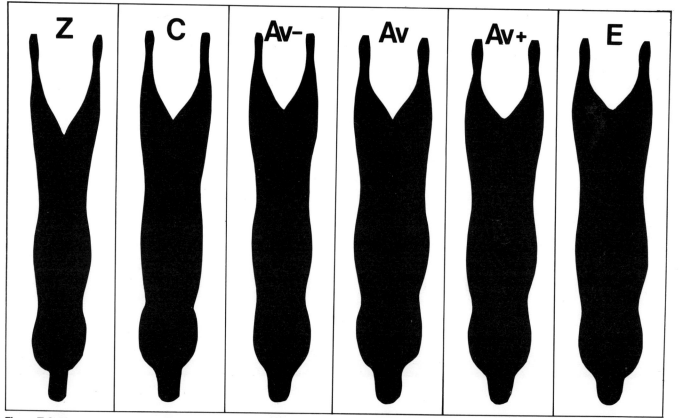

Figure 7.6 MLC Sheep Classification System – conformation

Table 7.1   The percentage of lambs of different fat classes

Fat class	Fat	Lean	Bone and waste
1	14.3	64.8	20.9
2	20.5	60.5	19.0
3	26.6	56.2	17.2
4	32.7	51.9	15.4
5	38.9	47.6	13.5

Estimating fatness in the live lamb

The farmer selecting lambs for slaughter must base his assessment on the live lamb. A system of condition scoring of lambs, similar to that used for ewes has again been devised by the MLC. The method relies upon feeling the back of the lamb in the lumbar region (point A in Fig. 7.7) to assess the depth of the muscle and fat cover over the backbone. In addition, the tail root or dock (B) is felt because this is the last part of the lamb to fatten and well reflects the fatness of the body. The shoulder and chest (C and D) are also checked for fatness. (See also Table 7.2.) The aim is to try to judge in the live lamb how the carcase will look after slaughter. To gain experience in this technique, it is helpful to visit the slaughterhouse to see the carcases of lambs that have been selected in this way and discuss with the meat trader the suitability of the lambs for his market.

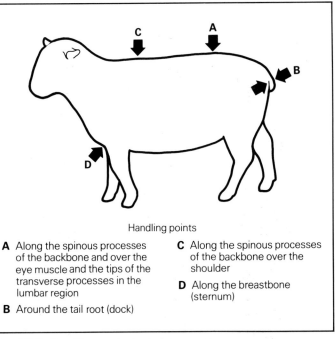

Handling points

A Along the spinous processes of the backbone and over the eye muscle and the tips of the transverse processes in the lumbar region

B Around the tail root (dock)

C Along the spinous processes of the backbone over the shoulder

D Along the breastbone (sternum)

Figure 7.7   Condition scoring lambs

It has been found that condition scoring of live lambs, when performed by a skilled operator, agrees very well with subsequent carcase classification. At Edinburgh recently, over 95 per cent of 600 lambs sent for slaughter at condition score 3 where classified as fat class 3 by MLC graders.

Table 7.2 Condition scores for lambs (from MLC (1978) *Sheep Facts*)

Fat class	Dock	Loin
1	Fat cover very thin. Individual bones very easy to detect.	Spinous processes very prominent. Individual processes felt very easily. Transverse processes prominent. Very easy to feel between each process.
2	Fat cover thin. Individual bones detected easily with light pressure.	Spinous processes prominent. Each process is felt easily. Transverse processes – each process felt easily.
3	Fat cover moderate. Individual bones detected with light pressure.	Spinous and transverse processes – tips rounded. With light pressure individual bones felt as corrugations.
4	Fat cover quite thick. Individual bones detected only with firm pressure.	Spinous processes – tips of individual bones felt as corrugations with moderate pressure. Transverse processes – tips detected only with firm pressure.
5	Fat cover thick. Individual bones cannot be detected even with firm pressure.	Spinous and transverse processes – individual bones cannot be detected even with firm pressure.

Differences between breeds

Breeds of sheep grow and fatten at different rates and reach different sizes at maturity. Estimated adult mature weights for different ram breeds are given in Table 7.3. In general, larger breeds like the Oxford and Suffolk grow faster and produce a heavier carcase at a given degree of fatness than smaller breeds like the Southdown or Dorset Down. If Suffolks and Dorset Downs are slaughtered at the same weight, the Dorsets will be fatter than the Suffolks.

Table 7.3 Estimated adult mature weights of sire breeds*

	kg
Southdown	61
Dorset Down	77
Texel	(89)
Suffolk	91
Oxford Down	107

(The weights shown are the average of males and females.)
*MLC (1978) *Sheep Facts* (figure in brackets based on few records).

These are not, however, universal rules. A notable exception is the Texel which is relatively slow growing compared to the Suffolk, for example, but is leaner at the same weight and reaches a mature size only slightly less than the Suffolk.

A light carcase (16–18 kg) with a moderate degree of fatness is produced by using a small ram breed like the Dorset Down.

A heavy carcase (23–25 kg) with the same degree of fatness is produced by using a Suffolk or an Oxford ram. A leaner carcase is produced either by using a Texel ram or by using a Suffolk, but slaughtering the lamb at a lighter weight (18–20 kg). The relative fatness of lambs at a given weight is also affected by nutrition.

The growth rates of crossbred lambs sired by different breeds of ram in a trial conducted by the Animal Breeding Research Organization are given in Table 7.4. The Oxford and Suffolk are the fastest growing of the breeds compared and the Texel the slowest. The Texel therefore took longer to reach the desired slaughter weight, but had a higher killing-out percentage (kg carcase/kg liveweight) and had a higher percentage of lean and lower percentage of fat in the carcase when slaughtered at 35 or 40 kg liveweight.

Carcase yields are given in Table 7.5 and carcase

Table 7.4 Growth rates of lambs sired by different breeds of ram on lowland ewes (data supplied by Dr C. Smith, ABRO)

	Liveweight gain 0–12 weeks (g/day)	12-week weight (kg)
Oxford	294	28.3
Suffolk	289	27.7
Oldenburg	279	26.9
Dorset Down	277	26.7
Ile de France	274	26.4
Texel	266	25.9

Table 7.5 Slaughter age and carcase production of lambs sired by different rams (mean of lambs slaughtered at 35 and 40 kg liveweight) (data supplied by Dr C. Smith, ABRO)

	Slaughter age (days)	Liveweight gain (0–slaughter) (g/day)	Killing-out (%)
Oxford	152	248	43
Suffolk	160	238	43
Dorset Down	173	218	44
Ile de France	175	218	44
Texel	173	216	44
Oldenburg	174	218	42

composition of the different breeds in Table 7.6.

Differences in growth rate between breeds can be explained on the basis of the efficiencies of feed utilization for fat and lean deposition. Because fat contains only 10 per cent water, whereas lean contains 70 per cent or more, fat has a much higher energy content than lean and an animal laying down fat grows more slowly on the same amount of feed because each kilogram of fat gain requires so much more energy. However, breeds also differ in the amount of feed they consume. The Texel appears to have a lower voluntary feed intake than the Suffolk and therefore grows more slowly and has less energy available for fat deposition because it consumes less feed.

Less information is available on ewe breeds because attention has been focused on their reproductive performance and milking ability rather than their growth potential.

However, growth and carcase characteristics of ewes are equally important because half the genes of the lamb come from its mother. In both crossbred and purebred flocks, at least 50 per cent of the lambs (the males) go for slaughter in any case.

Table 7.6 Carcase characteristics of lambs sired by different rams (data supplied by Dr C. Smith, ABRO)

Per cent lean		Per cent fat		Lean : bone ratio	
Texel	59	Texel	24	Texel	3.8
Oldenburg	56	Oldenburg	26	Dorset Down	3.6
Suffolk	55	Oxford	27	Ile de France	3.6
Oxford	55	Suffolk	28	Suffolk	3.4
Ile de France	54	Ile de France	29	Oldenburg	3.4
Dorset Down	53	Dorset Down	30	Oxford	3.4

The adult mature size of a breed is a guide to the optimum slaughter weight of the lambs. The larger breeds are likely to fatten at heavier weights and be faster growing than the smaller breeds. Table 7.7 gives the adult mature weights of ram and ewe breeds, estimated from records of mature weights of ewes and rams (the mid-parent value is used) by the MLC.

Optimum live slaughter weight is about half of the mature weight, thus to calculate the optimum slaughter weight of lambs, the MLC have suggested that the adult weight of the ram breed should be added to that of the ewe breed and divided by four. The predicted carcase weight will be approximately half the live slaughter weight of the lamb. The calculation assumes that an acceptable degree of fatness for the UK market is achieved at about 50 per cent of the mature weight and the killing-out percentage is about 50 per cent of the liveweight.

These assumptions are tenable when the lamb is grown at or near its maximum growth rate. Hence if a Suffolk ram (mature adult weight 91 kg) is crossed with a North Country Cheviot ewe (mature adult weight 77 kg), the mature weight of the crossbred lamb is 84 kg and the optimum slaughter weight

Table 7.7 Estimated adult mature bodyweight of some British breeds and crosses (MLC (1978) *Sheep Facts*)

		kg
Hill breeds	Welsh Mountain	45
	Cheviot	64
	Swaledale	64
	Scottish Blackface	70
	North Country Cheviot	77
Longwools	Romney Marsh	75
	Border Leicester	100
Shortwools	Clun Forest	73
	Dorset Horn	82
Crossbreds	Scottish Halfbred	89
	Welsh Halfbred	73
	Mule/Greyface	84
	Suffolk × Scottish Halfbred	89

(The weights shown are the average of males and females.)

is therefore 42 kg to give a carcase weight of 21 kg at fat class 3.

The mature adult weights shown in Table 7.7 are not necessarily the same as the particular weights of ewes and rams in an individual flock. Because of the differences in management and nutrition, individual flock weights cannot be used as a guide to the genetic potential of the sheep. The estimated adult mature weights of the breeds involved should be used to calculate the optimum slaughter weights of the lambs.

The calculated slaughter weight is the average for males and females. Lambs that grow more slowly or pass through a store period during growth can be taken to heavier weights before reaching the same degree of fatness.

The effects of nutrition

The rate of growth and the proportion of fat : lean laid down are both affected by the level of nutrition. On a high level of feeding, lambs grow faster and fatten at lighter weights than on a low level of feeding. Faster growth is accompanied by a greater rate of fat deposition in relation to lean. A lamb that grows fast will be slaughtered both earlier and at a lighter weight (at the same degree of fatness) than a slower growing contemporary.

The system of lamb finishing therefore influences the weight and fatness of the carcase produced. Lambs finished off the ewe's milk and good grass in summer will grow at near to their maximum potential and be slaughtered at 50 per cent of mature adult weight to give a carcase of nearly 50 per cent liveweight in the fat class 3. If lambs of the same breed are grown more slowly they will have to be killed at a higher weight to achieve the same level of fatness. If they are not fat by weaning and undergo a store period before finishing, overall growth rate will be less and heavier carcases will be produced. Thus, lambs that are finished off rape or Dutch turnips in autumn will be heavier than lambs finished off grass at the same degree of fatness, and similarly lambs that are not finished until winter or early spring, off swedes for example, will be heavier still.

Matching breeds to feeding and production systems

On a high level of feeding (intensive early lamb production or production from lowland grass), the fast-growing Suffolk or Oxford is used on lowland ewes to produce a carcase of 20–23 kg; or a Dorset Down is used to produce a smaller carcase (17–18 kg). The Texel is less suited to such systems because of its slower growth rate.

When lambs are grown more slowly to produce store lambs which can be finished at heavy weights, the Suffolk can again be used to produce a carcase of 22–25 kg which is not too fat,

Table 7.8 Weights of lambs (kg) sold off different production systems at condition score and fat class 3

	Suffolk × lowland ewes		Purebred Scottish Blackface	
	Liveweight (kg)	Carcase (kg)	Liveweight (kg)	Carcase (kg)
Intensive early lamb production	38	18	–	
Fat lambs sold off lowland grass	40	20	–	
Fat lambs sold off hill pasture	–		34	15
Autumn finishing on aftermaths, rape or Dutch turnips	45–48	21–22	36–38	16–17
Winter finishing on swedes, turnips, cabbages or indoors on barley	54	25	41	18

or the Texel used to produce a lean carcase of around 20 kg from hill ewes or 22–25 kg from lowland ewes.

Hill lambs grow more slowly because of the limitations of the environment, and moderately small hill breeds can be taken to 16–18 kg carcase weight without becoming too fat. If hill lambs are grown even more slowly and subsequently finished in the autumn on forage crops they produce carcases of 18–20 kg.

In practice the Suffolk is the most widely used breed for crossing with lowland ewes because it confers flexibility in time of finishing to suit many systems. It has the potential to grow rapidly to produce a 20 kg carcase off grass, but those lambs which grow more slowly may be finished later to produce a heavier carcase which still has an acceptable level of fatness.

The different weights of carcase that can be produced from two main types of lamb on different production systems are illustrated in Table 7.8 by the carcase weights of lambs sold throughout the year from the East of Scotland College farm at Bush, near Edinburgh.

Effects of sex

Entire male (ram) lambs grow faster than female (ewe) lambs. The performance of castrated males (wethers) is intermediate. Given the calculated optimum slaughter weight of a particular breed or cross, male lambs can be slaughtered 10 per cent above, wethers 5 per cent above and females 10 per cent below average to achieve the same degree of fatness.

Most male lambs destined for slaughter in the UK are castrated to avoid problems of mating behaviour when lambs are kept beyond 6 months of age. There is also trade resistance against uncastrated male lambs, but little evidence exists to suggest differences in eating quality. Because male lambs grow faster and are leaner at the same weight, there is a case for not castrating ram-lambs which will be fattened quickly for slaughter.

Growth promoters

Both male (androgens) and female (oestrogens) sex hormones promote growth. Both sexes naturally produce both types but in very different amounts. A balance of the two hormones is required for maximum effect. If additional exogenous hormones are given, females respond to androgens to give performance closer to that of the male, while entire males respond to extra oestrogens. Castrated males produce some androgens but no oestrogens, and therefore respond to the latter and possibly even more to a combination of the two.

Three artificial hormone-like substances are available which are claimed to promote increased growth. Hexoestrol is oestrogenic, trembolone acetate is androgenic and zeranol is only slightly oestrogenic. Only hexoestrol and zeranol are currently permissible for sheep in the UK. They are produced in small pellets which are implanted into the lamb's ear by means of a special gun.

Response to growth promoters depends on the nutrition and management of the lambs. Lambs on a high level of feeding with good management respond to implantation, whereas where nutrition or management are limiting a response is unlikely.

Trials on commercial farms have shown some response but this has usually been too small to justify the use of implants. There is also concern about the effects of growth promoters on secondary sexual characters of lambs. Hexoestrol is not recommended for ewe-lambs because it increases udder and teat development. Zeranol has been used to increase weights of ewe-lambs for early breeding, but it was found to have adverse effects on reproductive performance.

Nutrition of the weaned lamb

8

The very young lamb lives on milk and is adapted to digest milk in its true stomach (or abomasum). The rumen is small and milk can bypass the rumen by flowing along the oesophageal groove and straight into the abomasum. As the lamb gets bigger the rumen develops and it becomes adapted to digest grass and other solid feeds. Rumen development is accelerated by giving access to solid feed. Lambs can be weaned as early as 4–6 weeks for intensive early lamb production, but the traditional weaning age is 12–16 weeks for lambs off lowland grass or hill pastures. The time of weaning affects the development of the rumen – earlier weaning leads to earlier development – but the early weaned lamb is not *fully* ruminant until about 3 months of age.

As a ruminant the lamb relies on breakdown of the feed in the rumen by microbes which digest the feed. Sugars, starch and cellulose are converted into volatile fatty acids and protein and non-protein nitrogen into ammonia by the microbes. The ammonia is reconverted into microbial protein and the products all pass into the abomasum and the intestines for further digestion and absorption.

The advantages of this type of digestion are that the lamb can use roughage feeds which contain cellulose fibre and also simple nitrogen sources which can be converted into microbial protein. The disadvantages are that this method of digestion is not as efficient as the non-ruminant's and fibrous foods are still digested less well than starchy foods. Cereals are required for early weaned lambs and for maximum gains in older lambs. Early weaned lambs cannot use non-protein nitrogen and expensive proteins must be used.

Feeding the weaned lamb therefore involves a balance between the use of cereals and protein and the use of forages to achieve economic production. The diet of the lamb must contain energy, protein, non-protein nitrogen, minerals and vitamins in amounts that allow it to attain the required weight at the desired time.

Energy

Energy is supplied in different amounts by the food, but not all the energy is available to the lamb. Some energy is lost in the faeces after digestion; more is lost due to methane gas production in the rumen; still more is excreted in the urine. What remains is called *metabolizable energy* (ME) and this has been determined for different feeds (Table 8.1). It is measured in megajoules (MJ)/kg dry matter (DM). Cereals range from 12–14 and hays from 7 to 10 MJ/kg DM and the DM is about 85 per cent of the fresh weight. Grasses can have values as high as 12 for leafy material or as low as 9.5 for winter foggage and this is contained in a DM of about 20 per cent. Forage crops have very low DM contents ranging from 8–10 per cent for turnips and swedes to 14 per cent for rape, but the ME content of this DM is quite high at 9.5 for rape and up to 12.8 for swedes.

Table 8.1 The nutritive value of different feedstuffs

	DM (%)	ME (MJ/kg DM)	DCP (g/kg DM)
Barley	85	13	89
Oats	85	12	72
Wheat	85	13.5	105
Maize	85	14.2	78
Hay	85	7–10	30–90
Silage	18–36	7.5–10.5	98–116
Grass	18–23	9.5–12	100–185
Rape	14	9.5	144
Swedes	10	12.8	91
Turnips	8	11.2	73

The lamb needs energy both to maintain its body and to increase in weight. The amount of energy needed for maintenance increases as it gets bigger so that still more feed is needed to put on weight at higher liveweights. Furthermore, as the lamb grows it lays down progressively more fat in relation to lean. The energy content of fat is far greater than that of lean tissue so more energy is needed per kilogram gain as the lamb gets bigger. The capacity to consume feed also increases with the size of the lamb so it can maintain its growth rate for a time by eating more, but growth rate then slows down as the lamb reaches higher weight.

If the lamb is fed a high-energy diet to appetite, it grows fast. Less of the total feed is required for maintenance because it takes fewer days to reach the desired liveweight compared to a lamb on a lower energy intake. More feed is available for gain. If the amount of diet is restricted or the energy content of the diet reduced, there is less energy available for gain and the lamb grows more slowly, takes longer to reach the desired weight and a much higher proportion of the feed is consumed on maintenance.

Table 8.2 The effects of different dietary energy concentrations on liveweight gain and feed conversion efficiency of lambs fed to appetite between 20 and 40 kg liveweight (calculated from *Nutrient Allowances for Cattle and Sheep*, Scottish Agricultural Colleges, Pubn. No. 29, 1978)

ME content of the diet (MJ/kg DM)	Liveweight gain (g/day)	Time taken to increase from 20 to 40 kg (days)	Feed conversion ratio (kg DM/kg gain)
9	100	200	11.6 : 1
10	150	135	8.1 : 1
11	200	100	6.3 : 1
12	250	80	5.0 : 1
13	300	65	4.2 : 1
14	350	55	3.5 : 1

Table 8.2 shows how the amount of feed needed to increase in weight by a given amount (20 kg) increases progressively as the energy content of the diet and the rate of gain are reduced.

The lamb converts feed more efficiently on high-energy diets, but the economic advantage of this greater efficiency depends on the cost of the diet per unit of energy. A low liveweight gain and apparently poor feed conversion efficiency are acceptable if the feed is cheap.

The optimum rate of gain also depends on seasonal changes in the market price per kilogram. If the price is high (as in the spring), it pays to grow lambs fast so that they are sold before the price falls. If the price is low (as in the autumn), it pays to grow lambs slowly and accept a store period so that the lambs can be sold later, when the price has risen. A high-energy (cereal) diet is fed to early lambs which command a high price; a lower energy level is acceptable for store lambs in autumn and can be provided cheaply from forage crops.

Protein

Young lambs (when the rumen is not fully developed) cannot use non-protein nitrogen (NPN) but must be fed animal or vegetable protein to sustain high growth rates. Early weaned lambs grow fastest on cereal diets supplemented with fishmeal to provide 16 per cent crude protein in the ration. Soya-bean meal or a similar vegetable protein may be substituted for fishmeal but at a higher inclusion rate. The values of different protein sources are listed in Table 8.3.

Older lambs can use NPN sources such as urea because the rumen microbes synthesize protein. The microbes require energy as well as urea for protein synthesis and a high-energy (cereal) supplement must be fed with urea to allow effective

Table 8.3 The value of different protein sources

	DM (%)	DCP (g/kg DM)
White fishmeal	90	631
Linseed cake	90	286
Soya-bean meal	90	453
Groundnut cake	90	491
Urea	100	2200*

*Approx.

utilization. The microbes also require time for protein synthesis and, if a lot of urea is fed at one time, nitrogen is lost as ammonia and utilization reduced. The amount of urea that can be used is therefore limited and should represent no more than 1 per cent of the ration. Figure 8.1 illustrates the digestion of protein in the lamb.

Forage crops and grass are generally adequate in protein or NPN to supply the needs of slow-growing lambs and extra protein is required only when cereal diets are fed. This may be supplied by including 1 per cent urea in the ration which will elevate the protein content to an adequate level.

Minerals and vitamins

The diet must be balanced for the major minerals (calcium, phosphorus, magnesium and sodium), and trace elements (zinc, manganese, iron and cobalt and vitamins).

An imbalance in the mineral and vitamin supply can lead to reduced gains and deficiency symptoms. Cereal diets are low in calcium and if extra calcium is not given to bring the ratio of calcium : phosphorus up to 2 : 1 male lambs may suffer from urolithiasis (blockage of the urine passage). Ground limestone is therefore added to cereals for lamb finishing.

Cobalt deficiency is the most common trace-element problem and this usually occurs in store lambs on grass or forage crops in areas where the soil is deficient. Cobalt is required by the rumen microbes to synthesize vitamin B_{12} which is in turn required by the lamb. The deficiency can be prevented by injection of B_{12} but this is only effective for about 3–4 weeks. For lambs that will take some months to finish, a cobalt oxide bullet can be given; the bullet is given by mouth and rests in the rumen slowly releasing cobalt. Vitamins are required in small quantities, but again the main deficiencies occur with housed lambs on cereal feeds. A particular problem in intensive feeding has been vitamin E deficiency which causes muscular dystrophy in lambs. Grain preserved by treatment with propionic acid is particularly low in vitamin E. Muscular dystrophy is prevented by adding α-tocophorol (vitamin E) to the diet or can be treated by injecting selenium. Vitamin E and the mineral selenium are interchangeable in this respect.

When feeding cereals or cereal and hay diets to lambs, a powdered mineral and vitamin supplement is normally

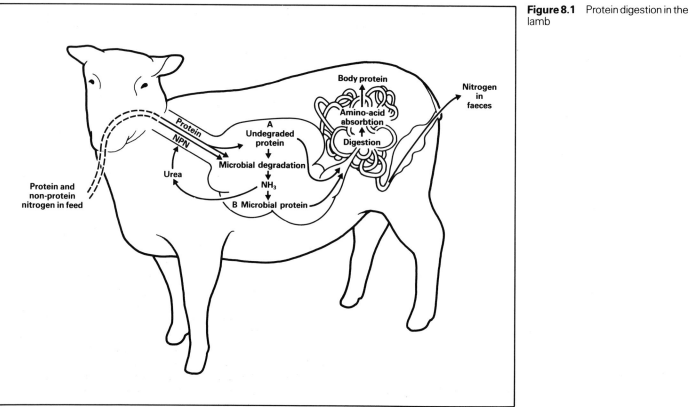

Figure 8.1   Protein digestion in the lamb

incorporated into the diet. Pasture and forage crops are normally adequate in minerals and vitamins except in certain areas where cobalt treatment may be required.

Feeds for lambs

Pasture is the natural feed for ruminants. Young leafy pasture provides a high energy level and is balanced with sufficient protein, minerals and vitamins to allow growth rates of 250 g/day or more. The nutritive value of grass falls if it is undergrazed and allowed to produce flower stalks. Autumn grass is invariably of lower quality than spring grass.

The inclusion of white clover in swards increases the protein and mineral content of the herbage and is believed to increase intake and utilization by lambs. All-clover swards are difficult to grow, have a short growing season and there is a danger of bloat if lambs are suddenly introduced to a clover field. Clover is therefore always grown in mixed swards in the UK.

In late summer, pasture growth declines and cannot provide sufficient feed for weaned lambs throughout the autumn. Other crops are therefore grown to feed lambs in the autumn and winter.

Forage crops

Kale, rape and Dutch turnips can be grown to supply feed in the autumn. Kale and rape are leafy crops which are rich in protein, minerals and vitamins. They can be fed without cereal supplementation, but cereals may be fed to increase liveweight gains or extend the use of a limited supply of the crop. They are low in fibre and a small amount of hay is sometimes fed to ensure proper rumen function. Kale and rape contain goitrogens which reduce the utilization of iodine but this is rarely a problem in practice. Feeding of an iodized mineral supplement with a low level of cereals can overcome the problem, but many lambs are fattened without with no deleterious effects.

Performance of lambs on kale is not as good as on rape, and rape is the better crop for lamb finishing. More recently Dutch turnips have been introduced as an alternative to rape. Both bulbs and leaf provide feed and the growth rates of lambs are usually as good as on rape. However, lambs do not fatten as readily on Dutch turnips unless some cereals are fed as a supplement.

All brassica crops contain substances which, when present in sufficient amounts, can cause anaemia in lambs. One particular chemical called S methyl-cysteine-sulphoxide (SMCO) has been identified as principally responsible. The content of SMCO varies with crops and with fertilizer usage. The content also increases as the crop matures and is highest in secondary leaf growth and flower. Kale and Dutch turnips are higher in SMCO than rape. Symptoms of anaemia are red-tinted urine and reduced appetite leading to poor performance. Problems are minimized if the crop is grazed before it is too mature and before secondary leaf growth occurs on a large scale.

Another problem with rape is the condition called yellowsis and occurs in white-faced lambs. It is due to factors which make the lamb sensitive to light and cause a skin condition on the face and ears. Rape is therefore not used for white-faced lambs, but the problem does not seem to occur with Dutch turnips.

Slow introduction of lambs to forage crops is important because the digestive system must adapt from a grass diet. Too rapid introduction to the crop leads to scour and deaths from 'brassica poisoning'. Gradual introduction can be effected either by limiting the period of access for the first week or by providing an adjacent area of grass which is eaten first before the lambs move into the crop. The grass area also provides a firm area for lying and for concentrate feeding to be carried out.

Turnips, swedes and cabbages

Although root crops have a higher water content, the energy value of the dry matter is nearly as high as cereals. They provide a useful and palatable feed in winter when other feeds are in short supply. Disease problems of the crop such as club root, finger and toe and mildew may lead to rotting and reduced total feed value as does frost damage.

Modern varieties of swedes far surpass the older varieties of both swedes and turnips for disease resistance and winter hardiness. The newer types tend to be harder because of higher DM contents but this does not reduce lamb performance. However, it does render them unsuitable for lambs which have lost teeth and causes more teeth loss in all lambs.

Cabbages have been used as an alternative to swedes, and modern varieties (particularly Dutch hybrids) are both high yielding and winter hardy. Teeth loss is significantly less than with swedes. The main problems with cabbages are the slow establishment of the seedlings and susceptibility to bird damage and weed competition.

Slicing or chopping of swedes for indoor feeding also reduces teeth problems and appears to give higher intakes and better performance. However, this detracts from the economies of outside fattening, requiring the use of a building and more labour input.

It is normal practice to feed a supplement of 250 g of cereals with winter crops and this may be increased if the crop is used up before all the lambs have been finished.

Hay and silage

Conserved grass products are variable in their nutritive value depending on the state of the grass when cut and the efficiency of conservation. Even good hay and silage are not sufficient alone for acceptable growth rates of weaned lambs. A combined ration of hay or silage and cereals can be used in preference to cereals alone for store lambs when a high rate of gain is not required. Hay requires supplementation with protein, but the higher protein content of silage means that cereals alone can be fed.

Rumen activity in the lamb depends on some fibre in the diet, and the feeding of some hay along with low-fibre diets is a method of ensuring sufficient fibre.

Cereals

Cereals are rich sources of energy and highly palatable to lambs. Maize, wheat and grain sorghum are the highest energy feeds followed by barley and oats. Oats have a thick husk and higher fibre content than the other cereals and the energy content is lower. The rumen digestion of cereals requires different microbes to roughage feeds and grass, and a period of adaptation is required when introducing cereal feeding. Too rapid introduction, particularly of barley and other high-energy feeds, causes acidosis. Affected lambs appear drunken and soon collapse. It is avoided by gradual introduction of 100 g/day for a few days then increasing to 200, then 400 and so on.

Cereals can be fed whole to lambs whereas they must be rolled or ground for cattle. The larger diameter of the gut in cattle allows cereal grains to pass undigested while this does not happen in lambs. Feeding processed cereals to lambs results in an undesirable pattern of rumen fermentation that produces propionic acid in the rumen. This results in deposition of soft fat which is not liked by the meat trade. The problem is overcome if a small quantity (10–15%) of roughage (hay or straw) is fed or if the cereals are fed whole. High-cereal diets can also cause rumenitis (clumping of the papillae which line the rumen) which does not occur if cereals are fed whole or with roughage. Furthermore, whole-grain feeding reduces the rate of rumen fermentation compared to that found with ground cereals and results in higher intake and digestibility of roughages when used as a supplement of forage feeds. Cereal processing is therefore both unnecessary and undesirable for lambs and oats, maize, wheat and barley should all be fed whole.

Barley is an extremely good feed for lambs but may require vitamin E supplementation if stored moist with propionic acid treatment. Oats give lower liveweight gains, and although usually cheaper are not as economical in terms of energy cost and performance compared to barley. Maize is the cereal with the highest energy content but it is expensive in the UK and has low Ca and Co contents. Wheat, because of the sticky nature of the gluten, can cause indigestion and it is generally recommended that it should not comprise more than 25 per cent of the diet. Grain sorghum can be fed, if available, as a component of lamb diets but causes constipation if fed in large quantities.

All the cereals are low in protein and calcium for fast-growing lambs. When cereals are fed as the main component of the diet, supplements of protein and minerals are required.

Table 8.4 Supplements for cereal diets fed to early weaned lambs (from *Feeding and Management of Early Weaned Lambs*, Scottish Agricultural Colleges, Technical Note No. 7, 1975)

A. To be included in the diet at the rate of 10 per cent of the total:

	(%)
Fishmeal	80
Ground limestone	15
Molasses	4
Minerals/vitamins	1

B. To be included in the diet at the rate of 15 per cent of the total:

	(%)
Soya-bean meal	85
Ground limestone	10
Molasses	4
Minerals/vitamins	1

Trace minerals should be included so that the total diet contains per kg:

$$150 \text{ mg } ZnSO_4 7H_2O$$
$$80 \text{ mg } MnSO_4 4H_2O$$
$$200 \text{ mg } MgO$$
$$5 \text{ mg } CoSO_4 7H_2O$$
$$1 \text{ mg } KIO_3$$

Vitamins should be included so that the total diet contains per kg:

5 000 i.u. vitamin A
1 000 i.u. vitamin D
  20 i.u. vitamin E

For early weaned lambs fishmeal or soya-bean meal can be used and suitable supplements are shown in Table 8.4. These can be compounded into 4 mm pellets which mix readily with the grains and preclude selection.

Cereal diets for older lambs can be supplemented with 1 per cent of urea which is sufficient to allow the rumen microbes to produce the necessary microbial protein. Dr Ørskov of the Rowett Research Institute has shown how urea may be included in grains diets by adding a solution which can also contain minerals and vitamins. A suitable mixture is given in Table 8.5.

Table 8.5 A solution of urea, minerals and vitamins for adding to whole barley in a diet for finishing store lambs (Dr E. R. Ørskov, Rowett Research Organisation, Personal Communication)

	kg/tonne barley
Urea	10
Calcium chloride	12
Sodium chloride	3
Trace elements/vitamins	1.25
Sodium sulphate	0.4

The diet is prepared in an ordinary vertical grain mixer. The calcium chloride and salt are dissolved in 30 litres of warm (40–50 °C) water. Urea can then be dissolved in the same solution. This is then poured onto the grain while mixing. The minerals and vitamins are suspended in 3.75 litres of water and added. Sodium sulphate is then added separately as fine crystals to the moistened grain (avoiding precipitation of insoluble calcium sulphate).

The amount of water added amounts to 3 per cent, but no storage problems result because urea is an effective grain preservative. The resulting concentrate looks just like grain but has a DCP content of about 110 g/kg at a cost that is much lower than the equivalent protein provided from conventional sources.

An alternative method of feeding grain is to use a mixer which incorporates ground straw, protein, minerals and vitamins in a complete diet. This method allows cereals to be introduced gradually by starting with a 75 per cent straw inclusion, reducing to 50 per cent after a few days, then 25 per cent and finally 10–15 per cent for maximum gain.

Non-nutritional factors

Given the weight of the lamb, intake and nutritive value of the diet, it is possible to predict a liveweight gain on a given system. In many cases, particularly with lambs outdoors such as on forage crops, the predicted gain is not achieved. The performance of lambs can be reduced by factors such as disease, physical defects (such as teeth loss or feet problems), weather and stocking rate on the crop.

Chronic parasitism markedly reduces liveweight gain and feed conversion. Some gastro-intestinal disorders result in temporary or long-term damage to the gut wall and reduce performance long after clinical symptoms of disease have abated. Worms are the main problem with outdoor feeding and strategic use of an anthelmintic is required to ensure optimal performance. Indoor lambs, and some outdoors, may suffer from coccidiosis which causes black scour. Routine use of a coccidiostat by drenching, injection or in water is essential for early lambs indoors.

Chronic pneumonia is another disease which can cause prolonged reduction in performance due to impaired respiratory function and spasmodic coughing. Vaccines are available which are effective against certain strains of *Pasteurella* but prevention is not guaranteed. Affected lambs may be treated with long-acting penicillin, but some damage may have already occurred.

External parasites (lice, ticks and keds) may also cause stress and reduce performance. Routine dipping of lambs should be carried out at the start of the fattening period. All disease problems are best avoided by routine preventive measures.

A particular problem with lambs for winter finishing is the loss of teeth. The lamb loses its milk teeth from about 9 months of age onwards and these are replaced by the adult teeth starting at between 12 and 18 months. This loss is accelerated by feeding on certain diets, notably turnips, and performance can be reduced if teeth are lost before the lamb is finished. Lambs which have lost several of their incisor (front) teeth should be fed a soft ration based on cereals and hay as they are unlikely to do well on roots.

Weather has a noticeable effect on lamb performance. Lambs on a forage crop 'break' suffer especially in cold, wet and windy weather. A sheltered area of grass lie-back helps to sustain good performance in autumn and winter.

Finishing store lambs on forage crops

9

Lambs produced on hill farms or lowland grass which are not fat by weaning are sold store for finishing on another farm or fattened at home. The most common methods of finishing store lambs use grass aftermaths (following hay or silage production), crop by-products, green forages or root crops.

Financial success from store-lamb finishing depends on obtaining lambs at an economic price, minimizing losses, providing a cheap source of feed and marketing the lambs to obtain the maximum price. Where special crops are grown for this purpose, a high yield must be achieved to permit a high stocking rate which spreads the cost over a large number of lambs.

Different feeding systems and forage crops suit different farming systems, breeds and sizes of lamb. It is important to decide what type of lamb should be produced or bought and the desired weight and time of sale under the local market conditions.

Available feeds

The main forage crops available for lamb feeding are listed in Table 9.1 and the periods when these can be used are shown. They are broadly divided into three catagories: by-products or forage left after crop harvesting (grass or roots); catch crops sown after earlier production of conserved grass or cereals; main crops which occupy the land for the entire growing season.

Table 9.1 Forage crops available for autumn and winter lamb feeding

	Sowing date	Period of use	Lamb grazing days/ha
Grass aftermaths		July–Sept	1 000
Sugar-beet tops		Oct–Nov	1 500–2 000
Catch crops			
Rape	June–July	Oct–Dec	2 000–3 000
Dutch turnips after grass	June–July	Oct–Dec	2 500–4 000
Dutch turnips after cereals	Aug	Nov–Dec	1 000–2 000
Main crops			
Yellow turnips	May–June	Oct–Jan	3 500–4 500
Swede turnips	mid-Apr–May	Nov–Apr	5 000–10 000
Winter cabbages	mid-Apr	Oct–Feb	4 000–6 000

The number of lamb grazing days is the stocking rate (lambs/ha)×the number of days each lamb spends off the crop. If the carrying capacity is 5 000 lamb grazing days/ha, this may be utilized by 50 lambs for 100 days or 100 lambs for 50 days.

By-products allow opportunist grazing for short periods of material that might otherwise be wasted. Catch crops allow a second enterprise to use the same land during the year but production is limited by the short growing season; the crops mature quickly to allow grazing 3–4 months after sowing. Main crops allow greater production per hectare but are in direct competition with other crops which could be grown on the land such as cereals. By-products cost very little to the sheep enterprise as they are simply leftovers which would not otherwise be utilized. Grass aftermaths could be used by other livestock and some additional fertilizer is required following hay or silage removal. The costs of growing a catch crop must be set against the lamb enterprise, but these are low relative to alternative feedstuffs such as cereals or hay. Main crop forages and roots can be costed in the same way, but the opportunity cost of the land should also be taken into account. Thus, if growing 1 ha of roots results in the loss of 1 ha barley, the income forgone from barley can be regarded as a cost to the lamb enterprise. However, where root crops are regarded as a necessary break in intensive cereal production, they may be justified on the grounds that overall cereal yields are improved as a result of the break crop.

When a special crop is grown for lamb feeding, it is essential to spread the costs over as many lambs as possible. The objective must therefore be to aim for maximum crop production and use a high stocking rate to utilize the high crop yield effectively.

High forage crop yield

Higher forage crop yields mean greater lamb-carrying capacity and reduced forage costs per lamb. Maximum yield of rape is achieved by sowing in mid-June (under UK conditions) and giving 100–120 kg N fertilizer per hectare. A high-yielding variety should be chosen, but where clubroot disease is a problem, only one variety (Nevin) has any resistance.

Highest yields of Dutch turnips are also obtained by sowing in mid-June and giving 75–100 kg N/ha. Later sowing is inevitable when used as a catch crop after cereals and lower yields will result. Less nitrogen should be used (50 kg/ha) as the potential response is less. Sowing after cereals is possible only after an early harvest.

Swedes will yield best if sown in mid-April in early districts and preferably not later than early May elsewhere. Recommended fertilizer levels are 80–100 kg N, 100 kg P_2O_5 and 80 kg K_2O/ha. Yields of the newer varieties of swedes are considerably higher than older types; Ruta Øtofte is the most consistent high yielder available in the UK. Where the disease is a problem, a clubroot-resistant variety (Marian or Wilhelmsburger Sata Øtofte) should be chosen.

New hybrid varieties of cabbage have given DM yields as high or higher than swedes but the advantage is not consistent. Seed costs are very high and total forage costs are far higher than swedes. Cabbages must be sown early (mid-April) for maximum yield and similar fertilizer levels to swedes should be followed with a further top dressing of 100 kg N/ha in early July.

High stocking rate-related to crop yield

High yields can mean high stocking rates of lambs and reduced costs per lamb. Variable costs are currently (1980)

about £70 per hectare for rape or Dutch turnips and £100 per hectare for swedes; costs per lamb can vary from £1 to £3 or more depending on stocking rate and other inputs. A good rape crop can support 60 lambs/ha for 50 days (September–November) and produce liveweight gains of 150 g/day. An average crop will carry less (40–50), fatten lambs more slowly and require a greater input of supplementary feed. A good swede crop will fatten 150 lambs/ha in 10–12 weeks (November–March) and require minimal concentrate inputs (200 g/head daily) to achieve 100 g/day liveweight gain. This compares with average stocking rates of 70–100 lambs/ha on most farms.

An assessment of the crop yield must be made before lambs are purchased and grazing begins.

It is worth weighing a sample of the crop to gain a more objective measure of crop yield. A few samples should be taken over the field from a known area (1 m^2 or 1 m of drill) and the weight per hectare calculated. The DM yield will be approximately 14 per cent of the fresh yield for rape, 8 per cent for turnips and 10 per cent for swedes.

The lamb requires an allowance of between 3½ and 5 per cent of its liveweight in DM per day on the crop. The lower figure applies at low crop yields and the higher figure should be used for high crop yields because utilization is reduced at high stocking rates. A 30 kg lamb requires 1–1.5 kg DM/day. A crop of rape of 2 tonnes DM/ha will provide 2000 lamb grazing days (40 lambs/ha for 50 days) and a 5 tonnes DM/ha crop 3 300 lamb grazing days (60 lambs/ha for 50 days).

Feeding concentrates reduces the demand on the crop and increases the carrying capacity: 0.2 kg concentrates per day will allow an extra 200–300 lamb days/ha.

The effect of crop yield on stock-carrying capacity of rape and swedes is shown in Tables 9.2 and 9.3.

Table 9.2   Autumn finishing of lambs on rape

Rape
Costs of production: £70 per hectare

Poor crop	Average crop	Good crop
2 tonnes DM/ha	3.25 tonnes DM/ha	5 tonnes DM/ha
2 000 lamb grazing days	2 800 lamb grazing days	3 300 lamb grazing days
40 lambs/ha for 50 days	56 lambs/ha for 50 days	66 lambs/ha for 50 days

Table 9.3   Winter finishing of lambs on swedes

Swedes
Costs of production: £100 per hectare

Poor crop	Average crop	Good crop
6 tonnes DM/ha	9 tonnes DM/ha	12 tonnes DM/ha*
5 000 lamb grazing days	7 500 lamb grazing days	10 000 lamb grazing days
70 lambs/ha for 70 days	105 lambs/ha for 70 days	140 lambs/ha for 70 days

*Figures for DM include leaf and bulbs.

Grazing system

The use of an electric or conventional sheep net to graze the crop in strips improves utilization. Little advantage has been demonstrated for this method of grazing rape compared to conventional set stocking, but it may allow better utilization of Dutch turnips. Strip grazing is the normal practice with the winter use of turnips, cabbages and swedes. The fence is moved weekly during the grazing period.

Provision of a grass lie-back area is advantageous for all forage crop systems; it allows the lambs to adapt gradually from grass to forage at the start, provides a firm area to lie and a place for concentrate feeding if required.

Lamb performance

Growth rates of lambs on forage crops are somewhat unpredictable. They are influenced by the previous management of the lambs and by the non-nutritional factors mentioned in Chapter 8. The DM content of the crops can vary considerably with rainfall and intake and performance is reduced when DM is low.

Estimates of expected performance of lambs grazing forage crops are shown in Table 9.4; these are based on practical trials and survey data.

Fig. 9.4 Expected performance of lambs on forage crops

	Daily liveweight gain (g/day)	Total gain during feeding period
Grass aftermaths	50–100	2.5–5 kg in 50 days
Rape	100–200	5–10 kg in 50 days
Dutch turnips	100–200	5–10 kg in 50 days
Swedes		
Turnips	70–150	5–10 kg in 70 days

Table 9.5 Appropriate finishing systems for different types of lamb

Lamb type	Weight gain to finish (kg)	System
Short keep	2–5	Aftermaths or by-products (rape or Dutch turnips)
Medium keep	5–10	Rape or Dutch turnips
Long keep	10–12	Swedes, turnips or cabbages

Breed differences may be expected with hill lambs growing more slowly and down lambs faster. Overstocking reduces the performance because there is an insufficient allowance of DM per lamb, but this can be overcome by introducing concentrate feeding. This will, however, increase the cost per lamb.

Types of lamb (see Table 9.5)

Short-term keep

At weaning some lambs require only to put on a little weight and condition before slaughter. These can be finished on aftermaths or arable by-products on lowland farms, or rape or Dutch turnips on hill farms. If these lambs are purchased rather than home produced, the price paid must be below the fat price for lambs or the equivalent weight as a short-term rise in the price of lamb at this time of year is unlikely.

Concentrates should be fed only if the fat price is well above the store price and forage feed is limited.

Medium-term keep

Lambs at weaning that require to increase by 5–10 kg in weight and 0.5–1 in condition score are suitable for finishing on rape or Dutch turnips. They are on stubbles or pasture until the crop is ready for grazing in September or October. Such lambs will finish on the crop in an average of 50 days and will therefore be marketable in November and December.

Long-term keep

Lambs which require to increase in weight by 10–12 kg and 1–1.5 in condition scores are unlikely to fatten on aftermaths, rape or Dutch turnips. They are kept cheaply on sparse grass or stubbles and may be used to clean up forage crops after

Figure 9.1   Lambs folded on swedes at Boghall farm near Edinburgh

early marketed lambs. Thereafter the lambs can be put onto swedes or other winter crops in December to finish in the New Year when prices are likely to increase (Fig. 9.1). Some concentrates (200 g/day) are normally fed to lambs on swedes, but this may need to be increased towards the end of the feeding period for lambs that have failed to finish on the crop. Thin lambs which may be missing teeth can be bought cheaply and are suitable for housing in the New Year for finishing entirely on barley indoors.

Although the long-keep lamb may be regarded as the poorest in appearance, they usually make the highest end price because they are sold in the New Year when the price of lamb is increasing towards its peak.

Buying price related to expected returns and costs

The main factors in profitability of lamb finishing are the buying and selling price of the lambs. Variable costs of growing the crop can be calculated from the cost of seed, fertilizer and sprays and divided by the number of lambs per hectare. Other costs include concentrates or other feeds, veterinary costs and miscellaneous extras. Interest on working capital for the period of fattening (2–3 months) should also be regarded as a cost. National forecasts or guaranteed prices can be used to estimate the expected price of the lamb when sold fat and, by deducting the costs and the profit margin which is

required, the buying price that can be afforded will be found. A reasonable target margin is £4.00–£5.00 per lamb; for medium- or long-keep lambs, the buyer can usually afford to pay £7.00–£8.00 per lamb less than he hopes to receive at the end. Short-keep lambs will cost less to feed but must be bought below the prevailing fat price with an allowance for a profit margin.

If the price in November is expected to be 135p/kg for a carcase weighing 24 kg, £32.40 will be received per head. Medium-keep lambs can be bought at around £25 in September to leave a margin of £5 when finished on rape. This is only one example, and a calculation of this kind is required for individual circumstances and types of lamb. The shrewd buyer knows his costs of production and has an idea of the end price before he buys his lambs. Table 9.6 shows how the farmers who make the most profit are those who buy lambs for less and sell them for more than the average.

Effective health control

Healthy lambs should be bought, preferably from a known source, and tick and fluke problems avoided if possible. Lambs should not be subjected to stress on arrival or starved before dosing. A worm drench should be given a few days later, using a modern anthelmintic which controls lungworm as well as stomach worms. They should also receive a pulpy

Table 9.6 Financial results of store lamb finishing enterprises (MLC (1978) *Sheep Facts*)

	Average (£)	Top⅓ (£)
Lamb sales	24.69	26.25
Less		
Lamb purchase price	20.32	19.71
Output	4.37	6.54
Variable costs	1.83	1.90
Gross margin/lamb	2.54	4.64
Gross margin less interest on working capital (at 10%)	2.00	4.15

kidney injection on arrival and again 2–3 weeks later, according to manufacturers' instructions. This disease is still the main killer of store lambs despite effective control methods. Abrupt changes in diet and management should be avoided throughout; these can precipitate death (from pneumonia) and, if problems are encountered, an early diagnosis should be sought and lambs run off onto old pasture for 24 hours or until the problem subsides.

Losses in store lamb finishing should be no more than 1–2 per cent, but if an effective health programme is not implemented deaths will soon rapidly deplete the profit margin.

Intensive sheep production and cereal finishing

10

In traditional sheep systems it takes at least 20 kg feed to produce 1 kg meat, compared to less than 5 kg feed to produce 1 kg bacon in pig production. The main reason for the difference is that the ewe produces about 1.5 lambs/year, whereas the sow produces about 18 pigs. But pig production relies upon cereal feeding, whereas sheep can use pasture and forages with low production costs from land with limited alternative uses. Because sheep are so much less efficient than other livestock, they cannot compete for the use of cereals as a major feed component.

The use of a limited amount of cereals can be justified only to allow acceptable levels of performance to be maintained at times when pasture production is limiting – as in winter or under very extensive conditions. If output per ewe is increased by raising the reproductive performance above traditional levels, it is necessary to feed more cereals but the cost can be covered by the additional output. More meat is produced and better use is made of other resources such as land and labour. It is necessary to carry out a detailed economic appraisal of costs and returns before embarking on a system of more intensive ewe management.

The young lamb is a reasonably good converter of cereals into meat, requiring 7–8 kg feed/kg carcase meat. By keeping the ewe cheaply or more efficiently and early weaning the lambs to finish on cereals, lamb can be produced economically and faster than by traditional methods.

Cereal finishing of lambs means that they must be kept in

confinement either in a house (in temperate regions) or in an open feedlot (in dry areas). Efficient use of housing, labour and feed necessitates the finishing of lambs in batches to fully utilize the facilities. Controlled reproduction is an aid to batch production and the use of hormones to synchronize mating and increase ewe productivity is a technique which has application in intensive sheep production. The alternative is to purchase lambs of similar size and condition from a number of extensively managed flocks to make up batches for intensive finishing.

All the intensive systems of sheep production which involve a greater use of cereals are more likely to be adopted when the price of lamb is high relative to cereal costs.

Intensive early lamb production in the UK

Hill and lowland sheep production off grass and forage crops results in a peak of supply in summer, autumn and early winter. The spring, when lamb is in short supply, is therefore a time when prices are high; there is also a traditional demand for lamb at Easter. Hence, lambing at Christmas to produce lambs which can be marketed in the high-priced Easter market can justify the extra cost of producing lamb when grass is unavailable.

Lambing must be timed so that lambs growing at about 300 g/day will reach market weight in April. This means lambing in late December/early January and to achieve this ewes must be mated in early August. For most British breeds, this is just before the start of the normal breeding season and it is necessary to use a breed with an extended breeding season.

The Dorset Horn has a long breeding season so is commonly used for early lamb production. The Finnish Landrace × Dorset Horn produces more lambs than the pure Dorset Horn, but may be too prolific for traditional methods and the smaller triplet lambs will fatten too slowly and require a proportion to be artificially reared. At the Edinburgh School of Agriculture, the early lambing flock consists of quarter Finnish Landrace/three-quarters Dorset Horn ewes which are intermediate in prolificacy and still have an advantage over the pure Dorset.

Suffolk and Suffolk × Scottish Halfbred ewes are also noted for their relatively early breeding season and as excellent fat lamb mothers. These are more readily available than Finn–Dorsets and are commonly used for early lambing. Scottish Halfbred and Greyface draft ewes are also used, but these must be weaned early if coming from a spring lambing flock if they are to be suitable for mating again in August and it is advisable to mate them towards the end of the month to ensure high conception rates.

A fast-growing breed of ram is needed to ensure maximum lamb growth rates and the Suffolk has been used for this reason at Edinburgh. Suffolk cross lambs produce a carcase that is too heavy for some markets and Hampshire or Dorset Down rams may be used to produce lighter lambs.

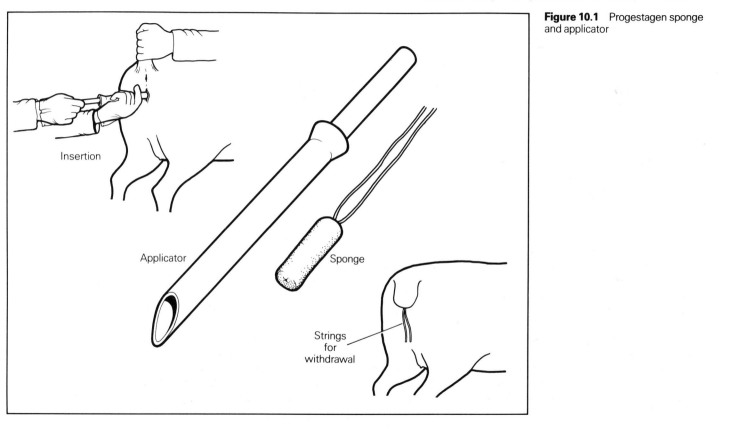

Figure 10.1 Progestagen sponge and applicator

Insertion

Applicator

Sponge

Strings
for
withdrawal

A compact lambing is desirable to simplify lambing management and to allow lambs to be batched for finishing. This is achieved by using hormone-impregnated sponges to synchronize mating.

The use of hormone sponges

Synchronized mating is achieved by suppressing oestrous activity (heat) for 14 days and then removing the suppressant so that all ewes come into heat at one time. Synthetic progestagens (called FGA and MAP) act in the same way as the hormone progesterone during the ewe's natural oestrous cycle. When their effect is withdrawn, the natural process of oestrus and ovulation is set in motion.

Synthetic progestagens are incorporated into small polyurethane sponge pessaries. A sponge is inserted into the vagina of the ewe by means of a special applicator (Fig. 10.1) and gradually releases the hormone into the body for 14 days. The sponge is then pulled out by means of the two nylon strings which are left protruding from the vagina. Oestrus occurs between 24 and 48 hours after withdrawal.

When ewes are synchronized, it is necessary to use a greater than normal ratio of rams to ewes – one ram to ten ewes. Conception rates at the first heat following synchronization during the breeding season are about 70 per cent, and the majority of ewes will conceive if allowed to mate at the next heat 17 days later, giving a total of 90 per cent conception at two heats. When synchronization has been attempted during the non-breeding season, conception rates have been below 50 per cent but are improved by injecting 500 i.u. Pregnant Mare's Serum (PMS) at sponge withdrawal.

For early lamb production, mating must take place at the very start of the breeding season. A conception rate of 90 per cent can be achieved with sponges without the use of PMS provided that mating is not attempted too early.

The lambing performance of the Edinburgh flock is shown in Table 10.1.

Table 10.1 Lambing results from the Edinburgh flock following synchronized mating in July/August

Ewes to ram	111	
Deaths	2	
Barren	7	
Lambed	102	(92%)
Total lambs born	198	(178%)
Lambs born alive	184	(166%)
Lambs weaned	174	(156%)

Management routine

Concentrate feeding normally begins 4–6 weeks before lambing and ewes are vaccinated against clostridial diseases at least 2 week before lambing. Ewes will continue to be fed concentrates and hay after lambing and preferably given turnips in addition. Lambs are weaned at 4–6 weeks of age and then fattened intensively indoors (Fig. 10.2). The ewes are fed a maintenance ration of hay indoors or out at grass.

A summary of the Edinburgh programme is shown in Table 10.2.

Table 10.2   Management programme for early lamb production

	Date		Day
Insert sponges	10	July	1
Withdraw	24	July	14
Mate ewes (1)	26	July	16
Repeat mating (2)	12	Aug	33
Commence concentrates (1)	20	Nov	133
Inject clostridial vaccine (1)	6	Dec	149
Commence concentrates (2)	6	Dec	149
Lamb (1)	20	Dec	163
Inject clostridial vaccine (2)	23	Dec	166
Lamb (2)	6	Jan	180
Wean lambs (1)	31	Jan	205
Wean lambs (2)	1	Feb	222
First lambs for sale*	Early Apr		
Average lambs for sale*	Early May		
Last lambs for sale*	Early June		

*At 40 kg liveweight (19 kg carcase.)
(1) indicates ewes mated at first heat after sponge withdrawal.
(2) indicates ewes mated again at second heat (17 days later).

Intensive feeding of early weaned lambs

The lambs are given access to a palatable creep feed from 7 to 10 days of age. A mixture of barley, flaked maize and soyabean meal is suitable. They are then weaned abruptly when at least 4 weeks of age; in practice an average weaning age of 6 weeks works best. The diet from then onwards consists of whole barley with a protein/mineral/vitamin supplement in 4 mm pellets mixed with the grain. This should be fed to appetite preferably from *ad lib.* hoppers or twice daily from troughs, ensuring that they are never without feed. Intake will increase from 0.6 kg to 1.5 kg over the fattening period. Lambs should be batched by date of birth in pens with approximately 50 per pen.

Growth rates will be about 300 g/day and lambs should reach slaughter weight of 35–40 kg in about 100 days. Because the feed is expensive they should be sold as soon as they reach marketable condition to avoid excessive feed usage. A total of about 100 kg feed will be required to take the lamb from weaning (15 kg) to 40 kg liveweight.

Management and health

Bad housing can lead to pneumonia problems and an allowance of 0.8m² per lamb is required with good draught-free ventilation. Feet problems may occur if they are not kept

dry with regular replacement of straw bedding or the lambs may be housed on slats or expanded metal flooring, which is better but more costly.

A black scour caused by coccidiosis is often encountered in early weaned lambs indoors and a coccidiostat should be given routinely by drench, injection or in drinking water.

Nutritional problems should not be encountered provided the diet is correctly balanced with minerals and vitamins and is available to appetite at all times. Of particular importance are inclusion of sufficient Ca to prevent urolithiasis in male lambs and vitamin E to prevent muscular dystrophy.

Acidiosis, caused by overeating barley, is likely to occur only if feed is withheld for any reason and then reintroduced too quickly.

Intensive systems for producing lambs all year round

Attempts have been made to lamb sheep more than once a year. To lamb every 7–8 months, ewes must be mated within 2–3 months of lambing and this necessitates early weaning of lambs for intensive finishing. This enables the ewe's body condition to be improved before rebreeding.

The most successful results to date have been obtained at the Rowett Research Institute in Scotland by Dr John Robinson. Finnish Landrace × Dorset Horn ewes were subjected to artificial daylength patterns (to simulate autumn daylength change) and given hormone pessaries as described above. The results are given in Table 10.3.

Table 10.3 Per cent conception and litter size over five consecutive reproductive cycles in two flocks of Finn × Dorset Horn ewes bred at 7-month intervals using photostimulation and progestagen treatment for the induction and synchronization of oestrus respectively (J. J. Robinson, C. Fraser I. McHattie (1977) in *Sheep Nutrition and Management*, US Feed Grains Council)

| | Reproductive cycle number | | | | | Overall means |
	1	2	3	4	5	
Month of mating						
Flock 1	May	Dec	July	Feb	Sept	
Flock 2	Aug	Mar	Oct	May	Dec	
Mean % conception						
Flock 1	94	88	85	82	85	88
Flock 2	98	92	85	81	83	
Mean litter size						
Flock 1	2.3	1.9	2.1	2.0	2.3	2.1
Flock 2	2.1	2.0	2.0	2.1	2.1	

The result was the annual production of 3.3 lambs per ewe. Recent work at the Rowett has shown that similar results can be obtained without artificial daylength patterns but still using progestagen sponges.

The success of this system is undoubtedly due to careful

attention to feeding and body condition at all times, and particularly before and during mating. This is more difficult under less controlled conditions and results elsewhere have been less favourable.

A more feasible system in practice may be an 8-monthly cycle which allows greater time between mating and rebreeding. In the UK, the timing should be such that one mating occurs at the normal time (October–November) and the other two are out of season (July–August) and late winter (February–March). Synchronization with progestagen sponges is essential, although some flocks of Dorset Horns have been bred at these time without hormones. A prolific type of ewe is essential to produce sufficient lambs to balance the high costs of management and feeding, and the Finn × Dorset Horn holds the most promise for UK conditions.

The ideal system is probably to have two flocks lambing three times in 2 years, but with lambings 4 months apart so that ewes that fail to conceive in one flock can be mated again 4 months later with the second flock. The programme is outlined in Table 10.4.

Table 10.4   A programme for lambing three times in two years

| | Mating period | | | | | |
	1	2	3	4	5	6
Flock 1	Oct	July	Feb	Oct	July	Feb
Flock 2	July	Feb	Oct	July	Feb	Oct

A system at the Northern Ireland Agricultural Institute has achieved production of 2.4 lambs per ewe in this system with local breeds. This is probably the minimum that can be accepted if the extra costs and management requirements are to be justified. At present it is doubtful whether a frequent lambing system can produce an economic result in practice.

Frequent lambing systems depend on intensive finishing of lambs born out of season, although grass may be used to finish the spring lamb crop. The relative economic performance of the system must be compared with that of the traditional production of lambs from grass. A comparison of the output and costs of lambing three times in two years with 244 lambs sold per 100 ewes annually, with those of lambing once a year with 159 lambs sold per 100 ewes is shown in Table 10.5. There is no difference in the final profit per ewe, but an advantage in profit per hectare because the intensive system uses a smaller total area. The system could only be justified on the basis of this calculation if grass area was the limiting resource.

There may be a place for frequent lambing systems in arid areas where cereal fattening is normally practised or where land is limiting and a high output is required from small flocks. Comparison of the economics of intensive systems with traditional systems must be carried out in the particular circumstances and it must not be assumed that increased output will produce greater financial returns.

Table 10.5 A comparison of the costs and output/100 ewes of lambing three times in 2 years with lambing once a year under UK conditions (from *A Study of High Lamb Output Production Systems*, Scottish Agricultural Colleges, Technical Note No. 16, 1977)

	Once a year 159 lambs sold/100 ewes	Three times in 2 years 244 lambs sold/100 ewes
Output	(£)	(£)
Lamb sales	3556	5620
Wool	280	280
Less replacement costs	−450	−670
Total output	3386	5230
Variable costs		
Hay	349	443
Concentrates – ewes	392	632
Concentrates – lambs	64	1250
Veterinary/medical miscellaneous costs	229	434
Total variable costs	1035	2759
Gross margin/100 ewes (excluding grass costs)	2351	2471
Interest (15%) on capital	390	491
Net margin/100 ewes (excluding grass costs)	1961	1980
Area of grass required	(8 ha)	(6 ha)
Net margin per forage ha	204	327

Intensive finishing of weaned lambs from extensive flocks

In arid areas of the world, poor, extensive pastures are used for sheep production. Nomadic systems enable sheep to be grazed over wide areas to use the scarce resources available. Production is limited to less than one lamb per ewe, but this is without supplementary feed unless given on an opportunity basis (arable by-products or stubbles).

In traditional systems, lambs are grazed along with the ewes and this results in poor growth rates and usually 2 or 3 years before sheep can be marketed or else very low carcase weights. Because both ewes and lambs are competing for scarce grazing, the range lands are overstocked in many countries where demands for food are increasing and output of traditional systems is low.

A better system has recently been adopted in Spain and certain Middle East countries which is modelled on the American feedlot methods. Range flocks produce lambs which are weaned earlier and put into large feedlots, often co-operatively owned, where they are fattened intensively. This reduces grazing pressure on the range, allows lambs to be finished faster and more efficiently, increases total output and therefore national food supplies and enables modern veterinary and nutritional techniques to be used.

The typical organization of a co-operative feedlot system is illustrated in Fig. 10.3.

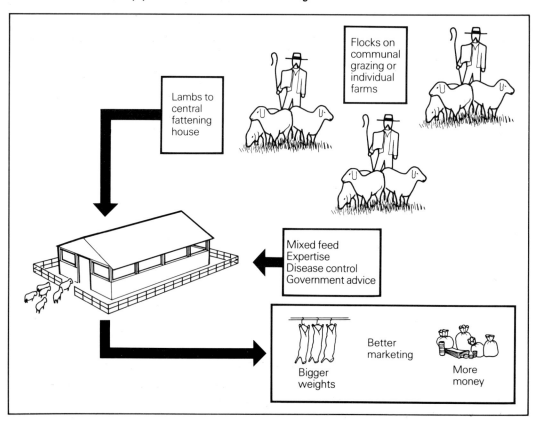

Lambs to central fattening house

Flocks on communal grazing or individual farms

Mixed feed
Expertise
Disease control
Government advice

Bigger weights

Better marketing

More money

Figure 10.3 Organisation of a cooperative feedlot (based on *Sheep Systems for Higher Meat Production,* US Feed Grains Council, 47 Upper Grosvenor St, London W1X 9PG, 1978)

The American feedlot system

Feedlots may be earth-floored enclosures in the open or houses with slatted or wire mesh floors. The ideal type depends on climate, shelter, numbers and available land. Lambs are fattened in batches on cereal-based diets designed to give maximum liveweight gain and feed conversion efficiency. The feedlot may be owned co-operatively by a group of flock-masters who finish their own lambs in this way or may be a separate unit which contracts to buy lambs from flock-owners or through a dealer. Production of lambs must be timed to coincide with demand for lambs by the feedlot, or the dealer must purchase lambs of a given weight and condition from different sources to make up a batch.

Large feedlots in the USA can finish up to 50 000 lambs at a time and feed 200 000 lambs annually. This allows efficient use of labour and equipment. They are usually sited at a strategic location in relation to lamb supply, feed sources and the slaughterhouse.

The management system

The lambs are weighed and identified on arrival at the feedlot. They are drenched for internal parasites, vaccinated against clostridial diseases, given vitamin injections and maybe a growth-stimulant implant.

Lambs are grouped according to weight and condition, so that similar lambs are treated according to their similar requirements and reach market weight at about the same time, enabling pens to be used and cleared at the same time with no need to sort lambs. In arid areas, lambs are sometimes shorn, depending on the season, wool values and condition of the lambs. Shorn lambs are less affected by hot conditions and can be stocked more intensively. Spraying or dipping is carried out to control external parasites and disease.

Feeding

Lambs taken off the range into the feedlot must be introduced slowly to a concentrate ration. They are normally introduced first to a high roughage diet (hay or straw). Then 25 per cent grain is incorporated for a few days, increased to 50 per cent, then 75 per cent and finally an 80–90 per cent concentrate ration (with limited roughage). Alternatively, they are fed a high roughage diet with grain fed separately, increasing from 100 g/day to 200, 400, 600 and so on until they reach appetite after about 2 weeks, while the roughage is gradually reduced to a maximum of 200 g/day.

Intensive finishing of store lambs in the UK

When suitable forage crops are unavailable and lambs are to be finished rather than sold as store, some lambs are also finished on cereal diets inside in the UK. Because of the relatively high feed costs of an indoor system, the aim is to sell lambs fat when prices are highest. Ideally, therefore, lambs should not be housed before December to sell them fat from mid-January onwards as the price rises. Those sold as late as March or April usually make the highest price per kilogram.

The management is basically the same as the American feedlot with concentrates gradually introduced as the lambs come off grass. Growth rates of 200–250 g/day are achieved so that 10 kg liveweight gain is obtained in 40–50 days. A recent innovation has been the use of urea-treated whole grain with added minerals and vitamins as the principal feed source, providing a cheaper diet than conventional concentrates.

The system is particularly suited to poorer, long-keep lambs which are not suitable for grazing on forage crops because of factors such as teeth loss.

Wool

11

Wool production was the *raison d'être* for the UK sheep industry 200 years ago. It was partly the demand for wool for industry which encouraged the Highland Clearances in Scotland when landowners began to favour sheep farming in preference to renting land to crofters. Today, with the advent of synthetic fibres, the importance of wool has declined to the position of a by-product of meat–lamb production. It is still the main product of the Australian and South American sheep industries, but in Europe, Asia and New Zealand meat is the primary product.

However, the takeover of the textile business by synthetics has not progressed as far as was predicted by the pundits of 30 years ago. The discerning consumer appreciates the superior qualities of strength, durability and insulation afforded by wool which no man-made fibre has yet succeeded in emulating.

Importance to the farmer

The relative value to the UK farmer of wool compared to lamb and draft ewes is shown in Fig. 11.1. It is clear that the lowland farmer derives very little from the wool clip relative to the sale of lambs, but wool from hill sheep still contributes about 15 per cent of their total output and is more important in that sector. Furthermore, the small differentials in price between grades encourage farmers to pay less attention to

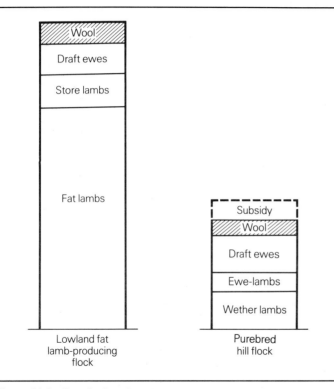

wool quality than to lamb growth and carcase quality. Few farmers consider shearing more than a necessary chore which contributes something to offset the feed costs of the ewe flock.

However, the effects of bad treatment and handling of wool – dirty, tinted or wrongly prepared fleeces – are easily avoided and correct handling ensures that the maximum price is obtained for the type of wool produced.

Weight of wool produced

By far the greatest factor affecting the farmer's income from wool is the weight produced. This varies between breeds and Table 11.1 shows the typical fleece weights of a number of British breeds. Under-nutrition can reduce fleece growth although wool continues to grow even when the animal is losing weight. British sheep are more susceptible to the effects of under-nutrition on wool growth than the Australian Merino which has been highly selected to produce wool under range conditions. Pregnancy and lactation also influence wool growth through their demands on nutrients. A greater weight of wool is shorn at the first shearing than at subsequent shearings, but the first wool clip is reduced when the ewe is lambed at 12 months of age.

Improved winter and summer nutrition increases wool production from hill ewes, and supplementary feeding during pregnancy with the use of improved pasture for lactation benefits wool output as well as lamb production.

Table 11.1 Wool characteristics of British sheep breeds (source: *British Sheep Breeds–their Wool and Its Uses*, British Wool Marketing Board)

Breed	Mean fleece weight (kg)	Mean staple length (cm)	Fineness (Bradford count)
Devon longwool	6.0	30	32–36
Teeswater	6.0	30	40–48
Romney Marsh	4.0	16	48–56
Border Leicester	3.5	20	40–46
Suffolk	3.0	6	54–58
Dorset Down	2.5	6	56–58
Dorset Horn	2.5	9	54–58
Clun Forest	2.5	9	56–58
Scottish Blackface	2.5	27	28–32
Swaledale	2.0	25	28–32
North Country Cheviot	2.0	7	50–56
South Country Cheviot	2.0	10	50–56
Welsh Mountain	1.5	8	36–50

Wool quality

Wool is made up of two types of morphologically distinct fibres – primaries and secondaries, illustrated in Figure 11.2. Primary fibres are thicker and come from more complex skin follicles (having a sweat gland, sebaceous gland and erector muscle). Secondary fibres are simpler (with only a simple sebaceous gland) and a higher proportion of these results in a finer fleece. Thus, fineness is determined by the ratio of secondary to primary fibres (or S/P ratio).

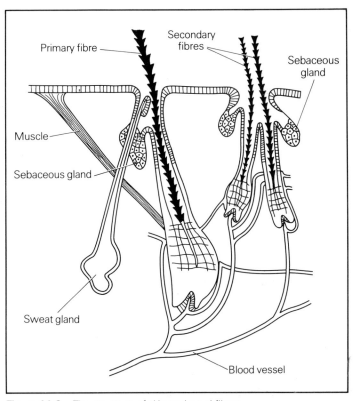

Figure 11.2 The structure of skin and wool fibres

In Merino, there are 15 – 23 secondaries : 1 primary while in British breeds the ratio is between 3 : 1 and 8 : 1. Primitive sheep also have a high proportion of 'kemp' (coarse fibres) and hair which are less desirable.

For grading purposes, wool is described by a 'Bradford count' which is supposed to be an estimate of the number of hanks, 510 m long, than can be spun on the Worsted system from 450 g of that wool. This is in fact determined subjectively by the grader. Merinos have 'counts' in the 80s while, at the other end, Blackfaces are in the range 28–30s. The qualities of wools of other British breeds are listed in Table 11.1, on page 123. Quality also varies in different parts of the fleece with the shoulders superior to the back and the neck and rump being coarsest.

Staple length is the measurement from skin to tip of a sample of the fleece. Shortwool breeds measure between 50 and 76 mm while longwools are between 300 and 500 mm (see Table 11.1). Length is important for Worsted and carpet manufacture.

White wool is preferred because it can be dyed to any colour, bright (or 'lustre') wool can be used white. Dark fibres detract from the value.

'Crimp' is the term given to the natural waviness or corrugations of wool which is responsible for its soft and springy texture. There are more crimps in fine wools – Merinos have 22–28 crimps per 25 mm while coarse wools may have as few as 8–10. Ryelands and Southdowns are the best UK breeds with 18–20 crimps per 25 mm (Bradford counts are in the 60s).

Wool prices and grading

Wool grading is practised by the buyers – the British Wool Marketing Board in the UK – to determine the price and to enable bulking of fleeces of similar quality and use. In the UK, the average price for the year is determined by agreement between the National Farmers Union and the Government at the Annual Price Review. The Board then deducts the estimated costs of handling and marketing and calculates the relative prices of different grades around this average. The list is published annually in the *Wool Price Schedule*.

Grading is based on type, fineness, length, colour and cleanliness. Examples of some of the different grades are given in Table 11.2 and definitions of some of the terms used are given in Table 11.3. Although many of the grade names are the same as breed names, grading is done by subjective assessment of the wool itself and a fleece may be given a grade which is not that of the breed of sheep from which it came.

The prices for different grades are determined by trade demand, but there are very small differences between the wool produced by a wide range of British breeds. The factors which have the biggest effect on price are cleanliness, colour and damage.

Table 11.2   Examples of British wool grades and prices (British Wool Marketing Board, ***Wool Price Schedule***, 1979)

Grade No.	Description	Price per kg greasy weight, p (to nearest 1p)
Down wools		
141	Merino	124
142	Merino cross	122
121	Dorset Horn	110
122	Dorset Horn Light Arable	101
123	Dorset Horn Arable	92
Fine wools		
203	Pick Teg	106
204	Pick Teg Light Arable	98
205	Pick Teg Arable	86
226	Super Ewe and Wether	111
227	Super Ewe and Wether No. 2	109
228	Super Ewe and Wether Light Arable	101
Medium wools		
338	Halfbred Ewe and Wether	115
339	Halfbred Ewe and Wether No. 2	113
340	Halfbred Ewe and Wether Light Arable	104
342	Deep Halfbred Ewe and Wether	113
343	Deep Halfbred Ewe and Wether No. 2	112
344	Deep Halfbred Ewe and Wether Light Arable	104
390	Halfbred Light Grey	92
391	Halfbred Dark Grey and Black	90
Lustre wools		
535	Roller Hog	113
Cheviot, Radnor and Welsh		
601	Cheviot Hog	114
607	Cheviot Hog Cast	108
608	Cheviot Hog Heavy Cast	95
Swaledale, Blackface and Herdwick		
722	Blackface Ewe and Wether Mattress	114
723	Blackface Medium Ewe and Wether	104
798	Blackface Hog, Ewe and Wether Discoloured	92
707	Swaledale	86
708	Swaledale Peaty	76
709	Swaledale Grey	75

Washing

A higher price is offered for washed wool of any grade. This involves dipping sheep in a flowing stream about 4 days before shearing. Not only does this clean the fleece but also makes shearing easier. It is still practised on some hill farms in the UK where suitable facilities exist, but is comparatively rare and the price advantage compared to the bother and loss in fleece weight makes it hardly worth while. It is still worth removing dirt from the fleece, particularly round the tail, before shearing and removing debris before rolling as dirty fleeces will be downgraded.

Table 11.3 Definition of terms used in British wool grading

Grade names	– This is the name given to the type of wool by the grader such as Cheviot or Halfbred, but these refer to the wool and not specifically to the breed or cross from which the wool came.
Pick	– fine wool with a staple length of 2.5–3.8 cm.
Super	– less fine than pick with a broader stronger staple of about 5–7.5 cm.
Halfbred	– less fine than super, about 10–12.5 cm long with broad, wavy staple.
Deep Halfbred	– about 2.5 cm longer, rather less fine and even broader staple.
Roller	– very long, lustre Hog wools with a sound, broad staple, 30–38 cm long.
Hog	– wool from the first shearing.
Ewe and Wether	– wool from the second and subsequent shearings.
Teg	– same as Hog but for fine wools only.
No. 2	– not very attractive or clean looking.
Light Arable	– earth or sand up to 6%.
Arable	– earth or sand more than 6%

Rise

Sweat from the sweat glands and grease from the sebaceous glands (see Fig. 11.2, p. 125 together produce the yellow 'rise'

in the fleece that helps shearing. The 'rise' usually occurs in May or early June in the UK and this is the best time to shear, given good weather. Shearing is normally a month or so later in hill areas.

It is possible to shear in winter before housing sheep, and trials at Aberdeen, Scotland, and Drayton Experimental Husbandry Farm, England, have shown the advantages of shearing prior to housing. Less respiratory stress was evident and a higher stocking density could be achieved, while some advantage in subsequent birthweights of lambs was demonstrated at Aberdeen.

Premature wool shedding

A problem frequently encountered in breeding flocks is wool shedding which normally occurs in late winter or early spring, just prior to lambing. It tends to be worse in some breeds, notably the Cheviot breeds and the Scottish Halfbred, but is mainly due to a nutritional check. Wool growth slows down in any case at this time of year, but a cessation of growth caused by a temporary shortage of feed (e.g. in bad weather or sickness) at a time of high demand for foetal growth results in constrictions in the fibres. These thin areas are liable to break and cause the wool to fall off. The remedy lies in adequate feeding, particularly of leaner sheep and during bad weather.

Shearing

Shearing is an art that cannot be learnt from books but only from practical experience. This is best gained by attending a training course of the type run by the Agricultural Training Board in the UK.

It is essential to use well-maintained equipment. In most developed countries hand shears have given way to mechanical devices usually driven by an electric motor. The handpiece consists of a comb and cutters driven by a gear mechanism via a flexible drive from the motor. Correct adjustment of the comb cutters is critical to good results and the handpiece should be regularly oiled during shearing.

When shearing the sheep, it is important to use long clean cuts or 'blows' to ensure that the wool is removed evenly and to avoid 'double-cutting' (when the fleece is cut part way through the staple as well as at skin level) which will result in downgrading. The sheep must be held so that the skin is taut in the area being shorn to avoid cutting the skin. At the time of year which sheep are shorn, small cuts attract flies and can lead to maggot infestation.

The sheep is caught from behind and lifted onto its rump on the shearing board. The belly wool is removed first and kept separate from the fleece. The order of shearing is then left rear flank, left side, left shoulder, head and neck, back (working in long sweeps from tail to neck), right shoulder and right side. The sheep is gradually turned round during shearing using the shearer's legs, knees and the left hand, which is also used to hold back the cut fleece. It arrives on its left side as the final blows remove the fleece from the right rear flank and then rolls onto its stomach and is released.

A good shearer can shear several hundred sheep in a day but, except in competitions, quality of work is more important than speed. Ideally, high quality should be combined with a high work rate.

Rolling the fleece

It is usual to work as a team with one or two shearers and one person rolling fleeces. The fleece should be thrown flat on a clean surface (Blackface hill breeds flesh side up, other breeds flesh down) and any debris removed. The sides are turned inwards and the fleece rolled towards the neck. The neck wool is drawn out and twisted to form a rope which is wrapped round the rolled fleece and tucked into itself. The fleece must be securely wrapped in this way to avoid a penalty.

The rolled fleeces should be packed into a woolsack. Dirty wool should be packed separately.

Penalties

As stated earlier, the flock-owner can ensure that he gets the maximum price for the type of wool he produces by avoiding the penalities that are incurred for marked or badly presented wool. These are listed in Table 11.4 at current prices.

The first few relate to coloration. Tinted dips, popular for show sheep but increasingly disappearing, should be avoided. Only scourable marking fluid should be used on sheep and no paints or other permanent marks.

Table 11.4   Deductions 1979

Artificially tinted fleeces	13p/kg less
Artificially stained fleeces	20p/kg less
Fleeces branded with tar, pitch or paint	10p/kg less
Fleeces tied with binder or baler twine	7p/kg less
Unwrapped fleeces	7p/kg less
Claggy and undocked fleeces	7p/kg less
Fleeces with daggs, locks of other oddments wrapped inside	7p/kg less

Conditions should be clean – a wooden floor for shearing is desirable. Straw and sawdust are particularly undesirable and should be avoided if sheep are housed overnight before shearing. Dirty wool should be removed from around the tail well before shearing.

The ability to bring sheep under cover overnight avoids them getting wet which can prevent or delay shearing the next day.

Wool must be correctly rolled, packed, labelled and stored in a dry place. Cleanliness and correct packaging greatly assists the marketing of wool.

Handling, housing and equipment

12

Well-designed equipment is essential to good sheep management. Labour costs have increased dramatically in recent years and this means that the shepherd must look after more sheep; this is made easier if he is provided with good handling pens which allow work to be carried out quickly and efficiently.

There is no ideal pen layout as requirements and sites will vary between individual farms, but the principles are common to all situations and can be adapted to suit the particular farm. If the pens are to give many years of useful service, considerable planning and forethought are required.

General principles

The site should be convenient for access both to the sheep and the shepherd and preferably near to water and electricity supplies. It should be free-draining and sheltered, but not overhung with trees which can shed slippery leaves onto working surfaces and encourage flies in summer.

Because sheep prefer to move up a gradient, a gentle slope is desirable with the direction of movement uphill, away from buildings and towards the hill or fields from which the sheep came. Long, narrow shapes and funnels encourage sheep to flow and they will readily move towards another sheep that they can see but can be distracted if they see sheep to the side of the direction in which they are required to move.

The layout should be as simple as possible, compatible with the work that must be done, and it should be possible to carry out the various tasks with the minimum of movement by the man. A continuous flow of sheep from the collecting area, up the race or into the working area should be the aim and it should be possible to recycle the sheep without disturbing other sheep in the working area or holding pens.

Gates and catches should be easily operable and all surfaces must be smooth, with no sharp edges. The ground surface must be non-slip and concrete is needed for the working area and draining pens for the dipper.

The choice of materials is between wood, metal and concrete blocks and the cost increases in that order. Wood is simplest to use but metal is stronger and requires less maintenance. Concrete is permanent but this makes it difficult to modify, enlarge or add to later. Recently, some well-designed, semi-portable metal equipment has become available that has proved efficient in practice. It is rather costly, especially for the larger layout, and a good compromise is to use this for the race, shedder and gates and to construct the holding pens and perimeter fencing from wood. All wooden posts must be pressure treated with a preservative before installation.

The sides of funnels and races should be sheeted, but gates at the end of a race should allow the sheep to see through so that they will move towards the light and other sheep at the end.

Facilities required

There must be a wide entry gate from the field into a pen which holds the maximum number that are regularly brought in, and a similar-sized area to turn them into after the work is completed. A working pen is needed for castration, tail trimming, foot paring and other physical treatment. A long race is needed with shedding gates that allow groups of sheep to be separated and put into different holding pens. It should be possible to stop sheep in the race to carry out dosing, marking, vaccination and condition scoring, or alternatively these operations may be done in the working pen.

A dipper is required for treating and preventing external parasites. This may be in the normal handling route or in a separate part of the layout. One school of thought is that sheep are more easily dipped if they are running in a direction to which they are accustomed; the other view is that it will be difficult to handle them for other purposes when they are afraid that they are about to be dipped. On farms where they are regularly brought in it is usually better to incorporate the dipper, but on hill units where the sheep are handled less frequently a separate dipper is preferred. A catching pen is needed at the entrance to the dipper and draining pens are required after dipping.

Additional facilities that may be included are a weigh-scale (which is portable and can be placed at the end of the race), a loading ramp and gate for getting stock onto vehicles for market and perhaps an integral shearing shed. Shearing can be carried out in a building that is used at other times for different purposes and this can be provided with a wooden floor, catching pens and a wool store for the shearing time.

Consideration may be given to putting a simple shelter over the working area. This gives protection in wet weather but can be hot in the summer or draughty in the winter.

Entry and collecting pen

The size depends on the number of sheep that must be handled at any one time; $0.5\,m^2$ is required per ewe and an additional $0.2\,m^2$ for lambs at foot. There should be a large entrance gate that can be fixed open. Ideally, the pen should be oblong with the exit in one corner, funnelling into a forcing pen.

Forcing pen

This may be a smaller rectangular pen with the exit at a corner, a funnel or a circular pen. A circular forcing pen is constructed with sheeted metal sides and a central post on which are hung two or three gates which extend to the full radius of the pen. Groups of sheep can be forced round from the entrance point to an exit at another point into the dipper, race or footbath. The forcing pen should hold a maximum of 20–30 ewes.

A useful feature for the funnel type is a two-way gate that can be swung inwards and then pulled through on rollers and brought back behind the sheep.

Working pen

Some people prefer to work in with the sheep while others like to vaccinate or dose ewes from outside. The latter can often be done in a race. A working pen (not a working race) should be about 1 m wide and 6 m long to hold about 10 ewes. This size allows the job to be done without the sheep getting past and, when finished, the treated sheep can be kept separate from the untreated.

A halfway gate is sometimes useful and it may be possible to have two pens side by side which allows one pen to be filled while working in the next. A shelf for equipment and medicines can be fitted to the side. Guillotine gates at the end, operated by rope and pulley, allow the shepherd to let sheep in or out from the other end. The floor of the pen should be free-draining and non-slip.

Race and shedder

This may be used for vaccination, dosing and marking as well as for sorting and shedding. It should have smooth, sheeted sides and sheep should be able to see out only through the exit gate. The sides can be vertical (0.375–0.500 m apart) or sloping

(0.450–0.675 m apart at the top and 0.200–0.300 m at the bottom) and about 0.875 m high or higher with a fold-down rail in one part to allow access for treatment. A gap at the bottom (0.075–0.100 m) provides a toe gap, can be used for checking feet, prevents young lambs suffocating and permits drainage.

The race may be used with a removable footbath or incorporate a long-swim dipper which is floored over when not in use.

The shedding arrangement at the furthest end may allow for two- or three-way shedding (see Fig. 12.1). Gates should be 0.900 m wide, and if two are used these should be staggered by 0.450 m to allow two-handed operation.

Footbath

Metal or fibreglass baths may be fitted into the race when required, or a separate permanent footbath built. There should be two stages – a wash followed by formalin treatment (5% formaldehyde solution). The wash-bath should be 3 m long and the treatment bath 6 m long with 0.075–0.100 m depth of liquid in both. The first 0.600 m can be flat-bottomed to encourage the sheep to walk in and the remainder is convex or corrugated to splay the cleats of the hooves. The side must be smooth and not have a ledge which allows the sheep to keep their feet clear of the liquid. It should be possible to hold the sheep in the bath by closing gates at both ends, and the exit

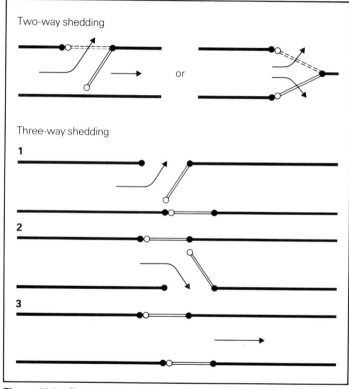

Figure 12.1 Shedding arrangements

should be onto a concrete draining area which may double as the draining pens for the dipper.

Dipping arrangements

The catching pen may be square, requiring no more than 2 m travel to catch the sheep, or it can be a circular forcing pen.

The size and type of dipper depends on the number of sheep to be dipped each time. The short-swim dipper is used for small flocks (200–500 ewes) and holds 900–1 300 litres. Only one sheep can be dipped at a time in a small bath and for larger flocks the long-swim type or circular swim dipper are more efficient. These hold 2 000–3 000 litres or more and therefore require the use of more of the insecticide chemical (dip).

The walk-in/walk-out type of long-swim dipper or the side-entry slipway on a short-swim or circular bath avoid the need to manhandle sheep into the bath. As an effective preventive of sheep scab, dipping must last for one minute per sheep and the circular bath has the advantage that they can be kept moving round for the prescribed time.

All three types of dipper are now available in fibreglass and this makes installation much simpler than with the older concrete construction. Examples of dippers are shown in Fig. 12.2, and Fig. 12.3 shows a circular dipper in use.

Two draining pens are best, each with the same capacity as the catching pen (0.5 m²/ewe), because it is necessary to let the sheep drip for about 10 minutes. The surface should be ridged concrete with a slope of at least 1 in 30 and channels so that the liquid runs back into the dipper via a filter pit.

An alternative to dipping is the use of a shower or sprayer. The sheep shower is a circular or rectangular tubular frame sheeted with corrugated iron. There are jets on pipes above and below the sheep which spray the insecticide. Although this method is effective against fly-strike, it is not approved for the compulsory dipping against sheep scab in the UK and has therefore declined in use.

Examples of layouts

Many of the above features have been incorporated into examples of sheep-handling systems shown in Fig. 12.4, 12.5 and 12.6. The sizes of collecting and dispersal pens and the type of dipper will vary with the size of the sheep flock for which they will be used.

Other handling aids

Other equipment is available that may assist the shepherd in his routine tasks. Foot trimming can be made easier by the use of a turning crate, a device in which the sheep is trapped, held and can be inverted exposing all four feet for treatment. A simpler arrangement is the sheep cradle which consists of two

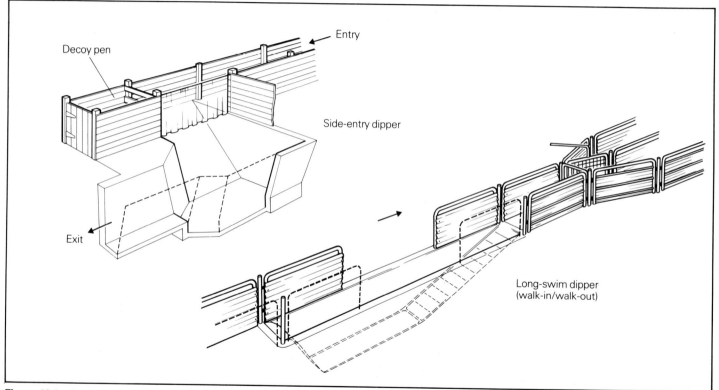

Figure 12.2 Examples of labour saving dippers

wooden poles with two cross-members loosely covered with a piece of canvas. The cradle is leant against the side of the working pen and the sheep is pulled backwards to sit in the frame.

Mechanical foot trimmers are also produced; these have an abrasive belt which is driven by a motor. The sheep is held in an elevated crate while its feet are mechanically trimmed from below. Most vets, however, consider this mechanical device an unsatisfactory method of treatment.

Sheep housing

Winter housing of sheep is practised to a small extent in the UK, but is an essential feature of management systems in other parts of Europe. In Scandinavian countries (Norway, Sweden and Iceland), sheep production is dependent on winter housing because of the severity of the climate. In the UK, winter housing is not essential, but advantages are claimed in certain situations.

The advantages

In wet areas, removal of the sheep from pasture in winter prevents poaching which delays spring growth and reduces

Figure 12.3 Circular swim dipper

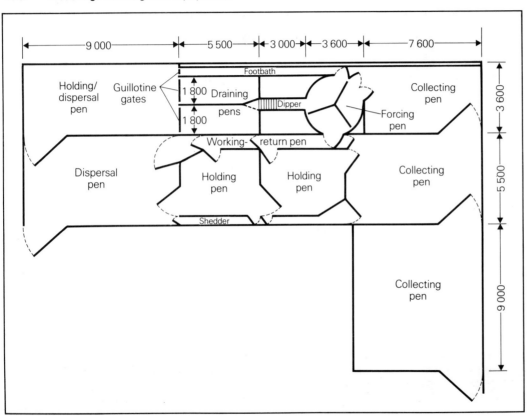

Figure 12.4   Sheep handling pens with short swim dipper

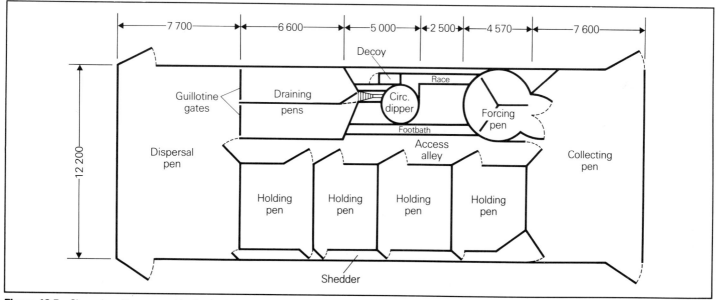

Figure 12.5   Sheep handling pens with circular dipper

total grass production. This allows higher stocking rates to be carried in summer. On arable farms, grass leys can be ploughed up for autumn sowing of cereals or to enable better soil weathering for spring sowing.

Working conditions for the shepherd are better, and closer attention to management is possible. Flock inspection is easier and it is possible to pen ewes in separate groups by age and condition score and to feed accordingly. Better supervision of

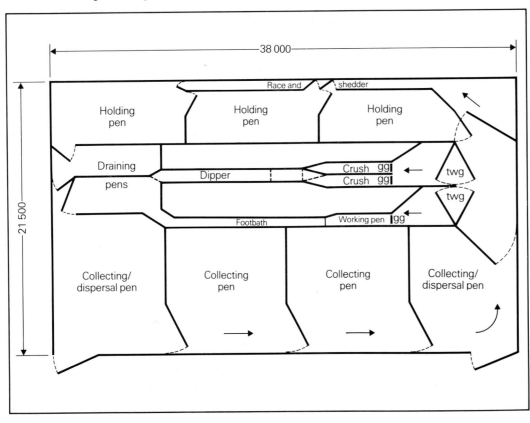

Figure 12.6 Large sheep handling pens (up to 800 ewes) with walk in/walk out dipper (gg = guillotine gate; twg = two-way gate)

lambing ewes can lead to a reduction in lamb losses and, together with improvements in feeding management, can result in a greater number of lambs produced. Older ewes will thrive better indoors and this allows them to be kept longer than in an outwintered flock, extending their working life and reducing depreciation costs. Mortality of ewes could also be reduced. Housing permits earlier lambing which enables more lambs to be sold fat at a high price.

In extensive systems, finishing lambs indoors relieves pressure on pasture for the breeding flock. Under UK conditions, it provides a system of finishing lambs when forage crops are unavailable or in hill areas where the winter is too severe for outside lamb finishing. It allows home wintering of immature hill breeding stock where this is otherwise impossible. Indoor wintering may be less costly than away wintering on rented lowland pasture.

The disadvantages

The high capital cost of a sheep house must be set against the profit from the flock. It is usually assumed for accounting purposes that the capital is repaid over 10 years and that the appropriate interest charge is added. There is also a cost of maintenance and repair of the building and equipment.

Although the energy requirements of housed sheep are less, total feed usage is usually higher because no contribution is derived from pasture. At Edinburgh, pregnant ewes kept indoors ate 2 kg hay/day compared to 1 kg eaten by ewes outside on a restricted grazing area. The extra feed adds to the costs each year.

There is a greater risk of the spread of infectious diseases, particularly pneumonia, coliform enteritis and foot rot, among housed sheep as well as increased levels of external parasites such as lice. Extra costs of preventive medicines are therefore incurred to avoid these problems where possible. Where hygiene standards are low, greater lamb losses may occur than in an outside flock and if ventilation is inadequate ewe deaths may be caused by pneumonia. Poor ventilation can also cause sweating and damp fleeces in finishing lambs and this may result in the transfer of taints to the carcase at slaughter.

Rubbing against walls and feeding troughs causes loss or reduction in value of wool.

Making the decision

The possible advantages of housing the sheep flock must be evaluated in relation to the disadvantages and especially the cost. For winter housing of pregnant ewes, each ewe requires 1.3 m² if bedded on straw or 0.93 m² on slats. Additional space is required for storage of hay and other feeds and for feeding troughs. The current (1980) minimum cost of a building is about £12 per square metre for a simple wooden-framed structure, but the cost could be as high as £40 per square metre for a steel portal frame. The additional cost of installing slats is about £15 per square metre. Feeding equipment and pen divisions will add more. Government grants are available in some countries and the current standard rate in the UK is 25 per cent. These costs vary widely between places and are subject to a high rate of inflation.

Table 12.1 shows a calculation of the annual cost per ewe of winter housing at current (1980) prices. In this example, a cheap wooden frame is erected by farm labour and the ewes are bedded on straw. Provision is made for fodder storage.

Table 12.2 shows the present profitability of the sheep flock

Table 12.1 An example of the calculation of annual housing costs

Do-it-yourself pole barn with fittings (Fig. 12.8)
Total cost for 500 ewes: £7500* (£15 per ewe)

	(£)
Capital repayment and interest over 10 years at 15%	1 500
Miscellaneous repairs, etc.	175
Total annual cost of building	1 675
Annual cost per ewe	3.35
Straw (5 bales or 75 kg/ewe at £25 per tonne)	1.80
Additional feeding (50 kg hay at £50 per tonne)	2.50
Extra veterinary and medical costs	0.17
Annual cost per ewe	7.82

*Net of 25% grant.

Table 12.2 Calculating the necessary increase in output to justify housing presents gross margin from out-wintered flock

Present gross margin from outwintered flock

Output	(£)
1.6 lambs at £28	45.00
0.2 draft ewes at £40	8.00
Wool	3.00
Less: 0.25 18 month ewes bought at £70	17.50
	38.50
Variable costs	
Concentrates	5.00
Hay	2.50
Grazing	5.00
Veterinary/medical and miscellaneous	1.00
	13.50
Gross margin per ewe	25.00

To cover additional housing costs of £7.80 per ewe
The number of lambs sold per ewe must increase from 1.6 to 1.9
      OR
Existing flock of 346 ewes must be increased to 500 because
      Total GM 346 ewes at £25 = £ 8 650
      Total GM 500 ewes at £25 = £12 500
    Less housing costs 500 ewes at £7.80 = £ 3 860
                 £ 8 640
      OR
Extra 77 tonnes of hay at £50 per tonne must be made on land not grazed by sheep in winter
      OR

Table 12.2 continued

Extra 48 tonnes of grain at £80 per tonne must be grown on land not grazed by sheep in winter
OR
Lambs sold per ewe increased to 1.75, an extra 30 ewes carried and an extra 20 tonnes of hay grown on land not grazed by sheep in winter.

(gross margin = output — variable costs) and the necessary increases in output that would justify housing the flock at these costs. Increases of the order shown in any one factor would be difficult to achieve but a combination of advantages could justify housing. The probable advantages must be estimated in the individual situation.

Although it is difficult to justify winter housing on economic grounds, it may still be felt that the benefits of easier management and staff comfort outweigh the costs. This is impossible to quantify and is a decision that must be made other than on economic grounds. It should also be noted that a large part of the recurrent costs are the feeding and bedding that are required over the winter period (3 months). Many of the management advantages of housing are obtained at lambing, and it may be better to consider housing for one week before and during the lambing period. This reduces the extra variable costs substantially and makes it more possible to use the building for other purposes such as fodder storage for the remainder of the winter.

To estimate the benefits of housing immature breeding stock on a hill farm, the building costs should be calculated as above (allowing $0.9\,m^2$/ewe hogg on straw, $0.6\,m^2$ on slats) and the costs of feeding, bedding and disease prevention added to give the total cost. The present costs of about £7 per head compare with the charges for away wintering in many areas and make this system attractive in these cases.

Fattening lambs can be regarded as a discrete enterprise and the profitability calculated in the usual way: output less costs. The purchase price of the lamb (or store value for home-produced lambs), feed costs, veterinary and medicine costs and miscellaneous charges for haulage, commission, etc. are added to the housing costs ($0.9\,m^2$/lamb) and this total is subtracted from the returns from finished lambs. The interest on working capital for the period of finishing must also be taken into account to find the net margin.

Reducing the costs

An elaborate, purpose-built house can rarely be economically justified. A simple building, made by farm labour from second-hand materials, may be a better prospect. Cheaper still can be the conversion of an existing building that is unused or a dual-purpose building used for crop or fodder storage at other times, thus spreading the cost over several enterprises. Outside yards are used in dry countries and have been tried in the UK, but climatic conditions are unfavourable.

Sheep require good ventilation, and the temperature need not be higher than outside provided they are kept free of draughts. They do not need an elaborate structure and a cheap, simple type can serve the purpose equally well.

Low-cost buildings

Many different designs have been used. Two examples are shown in Figures 12.7 and 12.8. Both are cheap and can be erected by farm labour from materials such as telegraph poles, corrugated metal sheeting and wooden paling which can be purchased second-hand. Straw bedding is used over a well-drained earth floor. Figure 12.7 shows a building suitable for a small flock of 150 ewes, whereas that shown in Fig. 12.8 is for a flock of 500 with provision for central storage of hay and straw. Sizes can be varied to suit the particular size of the flock. An example of a pole barn sheephouse, with ewes fed on silage, is shown in Figure 12.9.

The following features are essential. Good ventilation must be provided by leaving the front or sides open in sheltered locations or by the use of space-boarding on more exposed sites. Plastic netting or webbing has also been used as an alternative to wooden space-boarding, or corrugated iron sheets with 25 mm gaps between the sheets (every 750 mm) are another possibility. Draughts must be excluded at sheep level and the walls should be sheeted up to a height of 1.25 m. Care must be taken that draughts do not occur under gates or doors. In the case of a double pitch roof the ridge must be open along the entire length of the building. Another method of increasing ventilation is to have a slotted roof made from corrugated iron sheets with 25 mm gaps left between the sheets.

The floor area allowance is $1.3\,m^2$/ewe ($0.9\,m^2$ for lambs) and there is an allowance of 450 mm trough space per ewe for concentrate feeding. A much smaller amount of space is needed for feeding hay or silage which are available continuously. Trough space can be less for small ewes, but ewes with horns need more than a similar ewe without.

Group size depends on the most convenient division of the flock by age and condition, but very large groups can lead to crushing during feeding and handling operations – 25 per pen for small flocks and 50 per pen for large flocks are reasonable guides, but groups of 120 have been penned with no apparent problems. The shape of the building and the size and shape of the pens will determine whether it is necessary to have feed troughs on one, two, three or four sides. With small groups feed troughs may be placed in the pen, which saves space, but with large groups it is better to feed in troughs round the outside so that ewes eat through a barrier. Hay is provided either in racks above the feed trough or in combined hay/concentrate feeders which act as pen dividers (Fig. 12.10).

Slatted-floor lamb house

Another example of a building, for housing fattening lambs, is

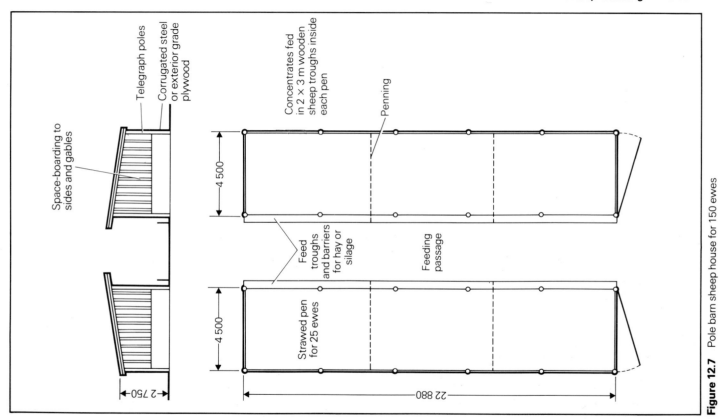

Space-boarding to sides and gables

Telegraph poles

Corrugated steel or exterior grade plywood

2 750

Concentrates fed in 2 × 3 m wooden sheep troughs inside each pen

Penning

Feed troughs and barriers for hay or silage

Feeding passage

Strawed pen for 25 ewes

4 500

4 500

22 880

Figure 12.7 Pole barn sheep house for 150 ewes

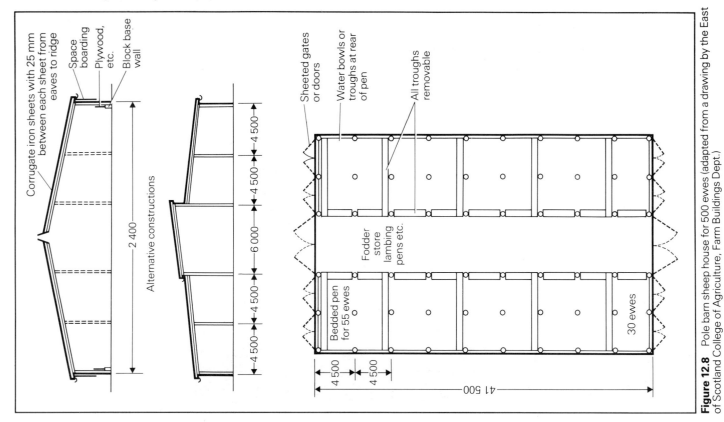

Figure 12.8 Pole barn sheep house for 500 ewes (adapted from a drawing by the East of Scotland College of Agriculture, Farm Buildings Dept.)

Corrugate iron sheets with 25 mm between each sheet from eaves to ridge

Space boarding

Plywood, etc.

Block base wall

2 400

Alternative constructions

4 500

4 500

4 500

6 000

4 500

4 500

Sheeted gates or doors

Water bowls or troughs at rear of pen

All troughs removable

Fodder store lambing pens etc.

Bedded pen for 55 ewes

30 ewes

4 500

4 500

41 500

Figure 12.9 Pole barn sheep
house with ewes fed on silage

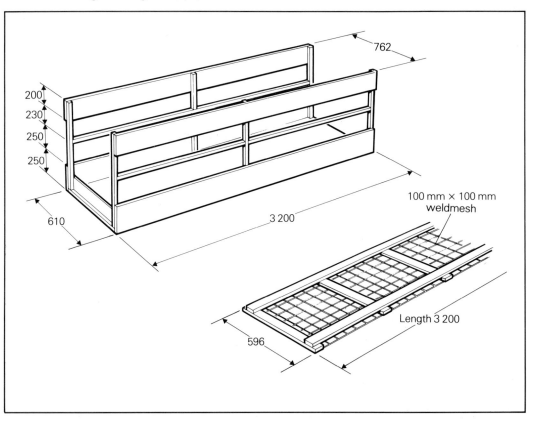

Figure 12.10 Portable combined feed trough/pen division for sheep

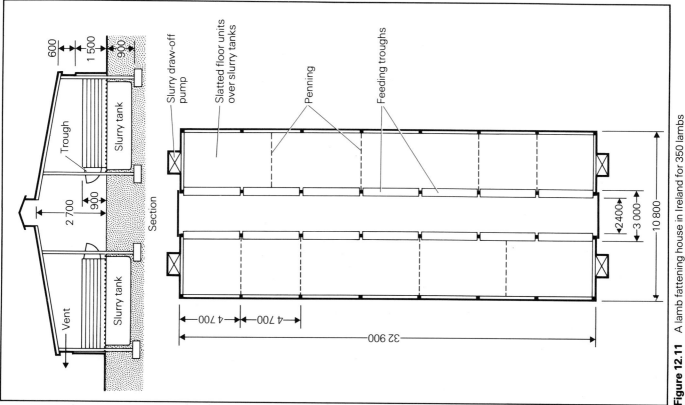

Figure 12.11   A lamb fattening house in Ireland for 350 lambs (From Sheep Systems for Higher Meat Production US Feed Grains Council, 47 Upper Grosvenor St., London W1X 9PG (1978))

shown in Fig. 12.11. This is a more elaborate design with a slatted floor to eliminate the need for straw bedding. An alternative to the wooden slats is expanded metal flooring which is popular in houses of this type in Northern Ireland. The liveweight gain of lambs has been superior to that of lambs on straw, cleaning out of dung was only occasionally needed and there was no labour requirement for upkeep. The occurrence of diseases such as foot rot and coccidiosis was greatly reduced. The area of floor required is less than on straw – 0.6 m²/lamb (1.0 m² is required for ewes on slats). Dung must be removed by mechanical or physical methods unless a deep cellar is used which involves extra material costs. Large troughs or hoppers are provided to allow *ad lib*. feeding.

Ventilation is effected by means of an open ridge with side vents above sheep level.

Avoiding health problems in housed sheep

The main hazards with housed sheep are pneumonia, foot problems and wool shedding. Pneumonia is unlikely to be a problem with good ventilation, but sheep may be vaccinated as a precaution. The effectiveness of currently available vaccines for *Pasteurella* pneumonia is *not* 100 per cent and they are effective only against certain specific strains of the disease organism.

Foot problems are greatly reduced by correct trimming of the feet before housing and regular (10–14 day intervals) use of a footbath containing 5 per cent formalin solution.

Wool shedding may be caused by nutritional stress, physical damage or parasites; it is important to ensure adequate nutrition, avoid sharp materials in construction and dip sheep 1–2 weeks before housing to eliminate lice.

An anthelmintic dose should be given 14 days after housing to remove stomach worms from previous grazing on infected pasture.

Sheep breeding

13

The objectives of the pedigree sheep breeder are to produce breeding stock for sale which are both attractive and have good performance, and to produce replacements for his own flock of equal or superior genetic merit. To the breeder selling sheep, reputation and future profit depend on the productivity of stock which are sold to other flocks. Breeding sheep may be sold to another breeding flock or to a commercial flock producing fat or store lambs. As flocks of the latter type make up the greater part of the industry, it is commercial production which must be the major concern of the breeder in determining the breeding objectives for his flock.

Unfortunately, it is the commercial traits which are the most difficult to assess in an individual sheep and often the most difficult to improve. In selecting individuals, the pedigree breeder very often concentrates on 'breed points' which are concerned mainly with uniformity of appearance and visual appeal and avoid the important attributes of reproduction and growth performance. On the other hand, the geneticist has little sympathy for 'breed points' and emphasizes selection for performance alone and usually aims for more or bigger lambs. But more and bigger lambs are not always economically desirable and do not suit all systems, particularly those where feed is limited (e.g. hill farms) or where small carcase size is required.

A modern breeder attempts to combine his stockmanship and knowledge of the industry with the use of objective recording and efficient genetic selection to achieve commercial success.

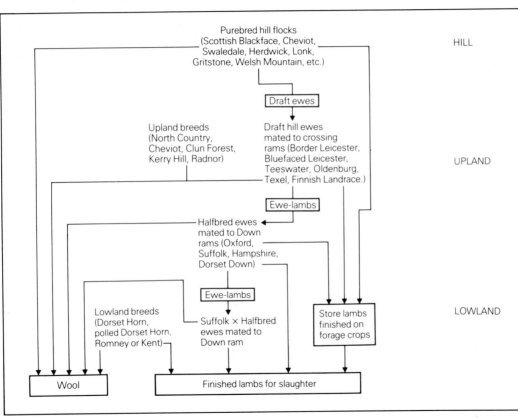

Figure 13.1 The stratification of the UK sheep industry

It is evident that reproductive performance and growth rate can only be assessed in a limited way by visual appraisal so that performance records are required to identify the best individuals. To be practical such records must be simple to obtain and be used in a breeding scheme designed to give the optimum response to selection. Some degree of visual appraisal of animals selected in this way is also required if the animals are to be sold in an auction and in order to avoid obvious physical defects of mouth, feet, body shape and fleece.

In the UK, there is a clearly stratified sheep industry (Fig. 13.1) with many different breeds and crosses to suit the many different systems and environments. Different importance is attached to the various physical and performance attributes and therefore different breeding objectives apply to hill breeds, crossing breeds (for producing lowland breeding stock) and terminal sire breeds (for crossing with lowland ewes to produce lambs for slaughter). For the improvement of a particular breed, the important production attributes need to be identified in order to permit efficient selection.

Breed substitution

To the commercial flock-owner, the choice frequently exists to change his current breed or cross for another. Normally, a faster rate of response can be expected by changing breed rather than selecting within the existing breed. Before embarking on an attempt at genetic improvement, the question must be asked whether another breed exists which already possesses the desired attributes. National recording schemes (such as that run by the MLC) are producing considerable data which enable breed comparisons to be made.

Table 13.1 shows the main attributes of some British breeds; it should not be regarded as a league table but rather a list of specifications from which the commercial producer should identify the breed or cross most suited to his particular system and conditions.

Hill breeds

The most important attributes of hill breeds are their ability to survive and produce a lamb under difficult conditions. Although weight of weaned lamb is a major determinant of profitability, it is affected as much by survival rate as by reproductive performance and growth. In New Zealand, much importance has been attached to easy-care sheep which produce lambs with the minimum of attention. However, efficient selection for this characteristic is probably better achieved by first identifying which physical characters are important for self-survival and easy lambing and selecting for these.

Table 13.1 Summary of performance of mature ewes in MLC recorded flocks (MLC (1978) *Sheep Facts*)

	Ewe bodyweight (kg)	Total lambs born per ewe	Live lambs born per ewe lambed
Hill breeds*			
Scottish Blackface	53	1.49	1.42
Swaledale	50	1.59	1.54
Welsh Mountain	35	1.16	1.13
Upland breeds			
North Country Cheviot	73	1.80	1.65
Clun Forest	62	1.72	1.65
Kerry Hill	61	1.60	1.42
Crossing breeds			
Bluefaced Leicester	86	2.25	2.01
Border Leicester	83	1.85	1.62
Oldenburg	65	1.71	1.52
Texel	82	1.75	1.52
Lowland breeds			
Dorset Horn	72	1.53	1.39
Polled Dorset	74	1.64	1.52
Down breeds			
Oxford Down	89	1.39	–
Suffolk	84	1.70	1.56
Hampshire Down	74	1.41	1.35
Dorset Down	74	1.41	1.33

*Performance of hill breeds was recorded in flocks kept in favourable environments.

Soundness of mouth, feet and legs and a protective fleece are of prime importance, together with good body conformation which is associated with a cover of fat as body reserves for winter survival. Lambing is likely to be easier in sheep with a fairly narrow shoulder but width in the pelvic region.

Because hill sheep are the source of much of the total supply of lamb, either directly or as a component of lowground crossbred ewes, fertility, growth and carcase quality cannot be ignored, but these can be conferred on the lowland crossbred by the use of an improved crossing ram on the hill ewe. Prolific sheep are subject to greater lamb mortality under difficult conditions so high prolificacy is undesirable in hill sheep; the ability to respond by showing increased reproductive performance when fed well in a good environment is valuable. Fast growth and lean carcase production is generally found in sheep with large mature size which could also be detrimental to sheep which must survive on limited feed resources. A type of fairly small mature size will fatten more easily, allowing some lambs to be fattened on hill farms as well as being able to lay down body reserves more readily in the ewe.

Wool characteristics are of relatively greater importance in hill sheep compared to lowland types because wool represents a higher proportion of the total income (15–20 per cent) and may also affect survival in bad weather.

Improvement of nutrition and environment can increase

Figure 13.2 A Scottish Blackface ram lamb

lamb production far more readily than genetic improvement. Scottish Blackface ewes, for example, can produce more than 1.5 lambs per ewe in the lowlands compared to less than 1.0 lambs per ewe in a hill environment.

Crossing breeds

The longwool breeds (Border Leicester and Bluefaced or Hexham Leicester) and some new exotic breeds (Texel, Oldenburg, Finnish Landrace and East Friesland) are used for crossing with hill breeds to produce ewe-lambs used as crossbred ewes in lowland grass flocks. Selection for prolificacy has been regarded as the principal objective for crossing breeds to improve lamb production in the lowland ewes. This is one of the more difficult characteristics to improve by selection, but progress has been made by selecting for performance over several years or several generations. Professor John Owen has developed the Cambridge breed which produces 2.6 lambs per ewe in the purebred flock by selecting ewes that produced three sets of triplets. Similar development of the ABRO Damline sheep by Dr Charles Smith and colleagues was based on selection within a multi-breed flock. The high prolificacy of these breeds requires high levels of management to realize their potential in purebred form, but when crossed with hill breeds they can come closer to achieving the desired optimum of 2.0 lambs per ewe.

Selection for growth and carcase traits has received less emphasis in crossing breeds with the notable exception of the Texel, which is less prolific but produces a leaner carcase. It must be remembered that all the wether lambs and about 80 per cent of ewe-lambs produced in crossbreeding flocks are slaughtered for meat. There is a strong case for greater attention to carcase characteristics in crossing breeds, although growth rate is less important because crossbred lambs are mainly produced on upland farms and sold store for finishing on forage crops.

A problem with the crossing breeds stems from the small flock sizes involved; a co-operative approach to selection would overcome this problem and also avoid inbreeding. Too much attention has been paid to fashionable breed points in the past and the commercial attributes have been neglected. A major departure from current practice is called for in the longwool types and considerable potential exists for development in this area.

Terminal sires

The Down breeds (Oxford, Suffolk, Hampshire, Dorset Down, etc.) are used mainly for crossing with lowland crossbred and purebred ewes to produce lambs for slaughter. Here the main objective is fast growth so that the maximum weight of lamb can be sold off grass. Recording and selection

Figure 13.3 A Border Leicester ram lamb

Figure 13.4   A Suffolk ram lamb

for growth rate is therefore required. Comparison of individuals under similar conditions in performance tests enables the breeder to select the best rams quickly and can provide data for advertising at sales (e.g. '150 day weight').

Other characteristics such as carcase leanness or conformation should be given greater emphasis in the earlier maturing breeds which are used for producing lighter lambs.

Rams selected on performance records may be judged visually to enable those with poor conformation, legs, feet, mouths or fleece colour to be culled. These secondary qualities cannot be neglected when making the final selection and especially when the rams are sold in the ring. Traditional judging must go hand in hand with objective improvement to ensure commercial success.

Genetics in sheep breeding

A given character may be controlled by a few or many genes. Simply inherited characters are controlled by a single or a few genes and the flock or breed is divided into very distinct groups: black and white sheep (fleece colour) for example. Black is the dominant gene in some breeds (e.g. Black Welsh Mountain) but recessive in most breeds. When black is recessive, only by mating black with black sheep are the offspring always black. Mating white sheep (which do not carry the black gene) to black sheep will always produce white lambs. However, these offspring will be carriers of the black

gene but because it is recessive it will be unseen. Mating two carrier (or heterozygous) white sheep results in three white to every one black lamb (see Fig. 13.5). If the breeder wishes to get rid of black fleece colour, he must not only avoid breeding from black sheep but must also keep family records and cull sheep which have black in the pedigree. In this way, the frequency of the black gene will fall to a low level in the population.

Other characters which are controlled by a few genes are jaw defects, presence of horns and lethal or semi-lethal defects such as dwarfism. In the case of horns, the character may also be sex limited – the ram only is horned (Welsh Mountain); other breeds are horned in both sexes (Scottish Blackface) or polled in both sexes (Suffolk). Furthermore, the inheritance of horns is also subject to incomplete dominance. For example, when the Derbyshire Gritstone (polled) is crossed with the Blackface (horned) the first cross offspring have small horns. It is necessary to select for some years in a crossbred population to obtain more or less completely polled sheep.

Most economically important characters (e.g. reproductive performance, growth rate, fleece weight, etc.) are controlled by many genes. The inheritance patterns are complex and selection for these traits is not as easy as for the simply inherited characteristics. Environmental factors (e.g. nutrition) also affect these characters and genetic variation is responsible for only part of the range in performance observed in the flock. At weaning a particular lamb may be large because of its own

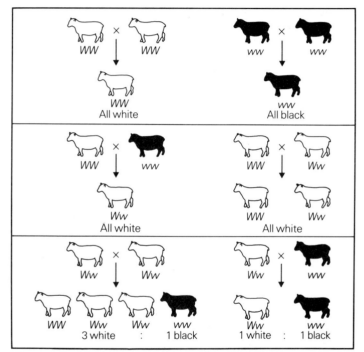

Figure 13.5 The inheritance of fleece colour (simple inheritance). Proportion of individuals having white and black fleeces resulting from a large number of matings. *W* = the gene for white fleece (dominant) *w* = the gene for black fleece (recessive)

growth potential or because it received more milk from its mother or grazed the best grass. The different extents of the influence of genetic and environmental factors determine the response that can be obtained by selection.

'Heritability' is a term used to estimate the proportion of the observed performance that is explained by genetics. The remainder is due to environmental factors such as nutrition, birth type, location, etc. Characters with a high heritability can be improved more easily by selection. A list of various performance traits and their estimated average heritabilities are given in Table 13.2.

Table 13.2   The heritabilities of some important traits in sheep (MLC, (1972) *Sheep Improvement*, Scientific Study Group Report)

Trait	Probable heritability
Mature body size	0.15–0.55
Lambs born per ewe lambing	0.10–0.20
Ewe milk yield	0.10–0.20
Lamb viability	0.00–0.05
Lamb growth rate	0.10–0.30
Carcase composition/conformation	0.25–0.35
Fleece weight	0.30–0.45
Fleece quality	0.40–0.70

Selection

If a graph is drawn of growth rates of lambs in a large flock or number of flocks, we find it has a typical shape (Fig. 13.6).

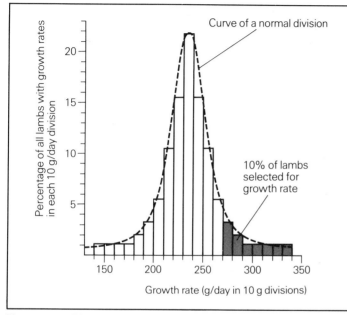

Figure 13.6 Distribution of growth rates in a flock

should be kept for breeding. The smaller the proportion selected and the nearer the top of the scale, the greater the response to selection that is achieved. The rate of progress is determined by the accuracy of selection and proportion selected, the heritability of the trait and how often selection can be carried out (the generation interval).

As a smaller number of rams are required to mate with a given number of ewes, a greater selection pressure can be applied to the rams than can be applied to the ewes. It may be possible to use only the best 1 per cent of rams, but we must accept the best 50 per cent of ewes. Although the ram side makes the biggest contribution to improvement, it is still essential to select on both ewes and rams to make effective progress. Selection of individuals in this way works well for growth rate or fleece weight which can be measured in the live lamb. It cannot be used for carcase quality or when the character is shown in only one sex, e.g. ewe prolificacy (sex limited). For these characters we need more extensive records and must select on the basis of the performance of the relatives.

This is called a 'normal distribution' and shows that there are more lambs around the average and fewer at the top and bottom. If an increase in growth rate is required, only lambs with high growth rates (at the right-hand side of the graph)

Pedigree selection

If objective information is recorded about individuals within a flock for a period of years, selection of an animal for breeding can be based on the performance of its ancestors. If wishing to select a ewe-lamb for prolificacy, the mother's record, maternal grandmother's record, paternal grandmother's record, etc. can be used. The mother contributes half the genes, the grandmothers a quarter and so on. The heritability of prolificacy is higher when performance is measured over several lamb crops. Hence, selecting a ewe-lamb whose mother and grandmothers had triplets three times each is more effective than simply using a lamb which was a triplet itself (one record).

Progeny and sib testing

Another technique is to test an animal by measuring the performance of its sons or daughters. The best estimate of a ram's breeding value is obtained by mating it to a large number of ewes and recording the performance of its offspring. The problems are that this takes longer and involves much greater expense (because a large number of animals must be kept for the test) than individual selection. It has been little used in sheep breeding although it is a common method in dairy cow selection where milk yield is sex limited.

The time taken for a progeny test is reduced if selection is based on the performance of brothers and sisters – a sib test – but this still requires a larger number of animals to select one individual than pedigree or individual selection. However, the accuracy of predicting breeding value is far greater in the progeny or sib test.

The situations where these techniques are being effectively used are in selecting for growth rate in Welsh Mountain rams and also in Suffolks and Cambridge rams where individual performance, progeny and sib testing have been used by Professor John Owen in association with the MLC.

Selecting for more than one character

When defining objectives for selection, two or more characters of importance may be identified. If this is the case, a selection index can be constructed to assign an animal a breeding value based on two or more aspects of its performance. Each character is given a mathematical weighting in respect of its relative economic importance. An index made up of two different records is calculated as:

$$I = W_A A + W_B \bar{B}$$

Where A and B are measured values of the two traits and W_A and W_B are the respective weighting factors.

In practice, the fewer the factors considered, the greater the progress in each because selecting for several characters at once reduces the selection pressure for all. It is vital that the objectives are carefully defined and as few factors as possible are involved. If it is essential to select for two characters, the highest weighting must be given to the most important commercial trait.

Genetic correlation

Another problem arises when the inheritance of two characters involves some of the same genes and therefore selection for one causes an inevitable change in the second. An example of this in sheep is the genetic correlation that exists between growth rate and mature size. Attempts to increase growth rate invariably result in an increase in mature size and therefore the weight of the breeding stock. If lamb growth rate is increased, lambs will have a higher carcase weight at a given degree of finish, and ewes will be bigger and therefore require more feed for maintenance. This has serious implications when feed is limiting as in the hill situation and where it is difficult to finish lambs from large breeds.

An alternative strategy is to select for 'relative growth rate', that is the growth rate of lambs relative to the size of its parents. The problems in adopting this apparently ideal character for selection are that it is difficult to measure and the variation in this character is believed to be small, making it difficult to find superior individuals.

In practice, growth rate should not be emphasized unless it is certain that other consequences of this decision are not detrimental to a particular breed in the given environment. It is of most importance in terminal sires like the Suffolk which are used on other breeds of ewe to impart a high growth rate to the offspring.

Problems of environmental effect

The environmental effect on performance can be minimized by attempting to standardize the environment during selection. Professor Owen used artificial rearing and pen feeding of Suffolk rams for performance and progeny testing. The Welsh Mountain progeny test was also carried out in a good lowland environment.

One problem is the possible interaction between genetics and the environment: a sheep may perform better than its contemporaries in one environment (e.g. in the lowlands) but worse in another (e.g. in the hills). The evidence for genotype/environment interactions in sheep is limited. It would seem better to select hill sheep under hill conditions, but in practice it is much easier to run a breeding flock in the lowlands. However, the possible consequences of a genotype/environment interaction should be borne in mind and any scheme should seek to validate its results at an early stage by some testing in the environment in which the sheep are to be used commercially.

Effects of age, sex and parity

It is important that comparisons are made between animals which are contemporary in respect of age, sex and, in the case of ewes, the number of lamb crops previously produced because these factors influence performance. Weighting factors can again be used in calculating the breeding index of an individual for whether the growth rate was measured on a ewe or a ram lamb and whether prolificacy was measured in one-crop, two-crop or three-crop ewes. The use of correction factors enables all available information to be used.

Inbreeding and crossbreeding

If close relatives are mated, performance of the offspring is inferior because they are said to suffer from 'inbreeding depression'. Conversely, the mating of totally unrelated strains or breeds results in an improvement in some aspects of performance above the average of the two parents. This latter effect is known as 'hybrid vigour'.

It is therefore important that the mating of related rams and ewes in a breeding programme is avoided. Recording of the pedigree of individuals within the flock makes it possible to mate a selected ram to an unrelated group of ewes. Individuals may be grouped by families and mating carried out on a cyclical basis so that a ram is mated to ewes of a different family each year (see Fig. 13.7). In flocks in which ewes are replaced by ewe-lambs from the same flock and detailed pedigree recording is not possible, rams should be purchased from other sources at regular intervals to avoid inbreeding. A ram of a totally different breed or strain may be used from time to time to introduce a measure of hybrid vigour. Swaledale rams are used in this way every 4–5 years in some Scottish Blackface flocks.

Inbreeding has been used in other species to 'fix' important characteristics of a few superior individuals. Inbred lines are created by mating offspring to one of the parents or to their half or full sibs. The inbred lines are then crossed to remove the effect of inbreeding depression. This technique increases the uniformity of the lines with respect to simply inherited characters such as polledness and fleece colour, but leads to depression in the quantitative characters of economic importance and no greater uniformity in these traits. The technique has little value as a method of increasing production in sheep breeding.

The use of crossbreeding of different sheep breeds is widely used especially in the UK for combining the attributes of the two types and also achieves a measure of hybrid vigour in the crossbred offspring. Border Leicester rams, from a breed noted for size and prolificacy, are mated to the hardy Scottish Blackface ewes to produce the very productive Greyface for use in the better uplands and lowlands.

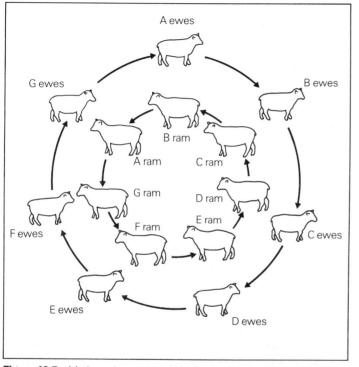

Figure 13.7 Mating scheme to avoid inbreeding in 7 families (after J. B. Owen (1976) *Sheep Production,* Bailliere Tindall)

Performance recording

Individual sheep must be identified (see Fig. 13.8) so that the pedigree is established and mating organized to avoid inbreeding. Accurate records must be kept of performance of the individual sheep. The records can then be used to select superior individuals for future use in the flock and as objective information for prospective purchasers.

For the last 10 years, the MLC have operated a national flock recording scheme in the UK designed to add performance data to normal pedigree information. The essential records required are those which will establish the dam and sire of every lamb born, the reproductive performance of the ewe and ram and the growth rate of the lamb. Ewes are weighed and condition scored at mating and the ram weights taken where possible; lists of matings, dates of lambing, sex, birth type (single or multiple) and birthweight (optional) of lambs are recorded; a 56–day lamb weight is obtained by weighing all the lambs in the tenth and thirteenth week after lambing commences; a 112–day lamb weight may also be taken (optional); fleece weights and fleece grades (optional) can also be recorded; records are kept of deaths and disposal of all animals.

The records are processed centrally by computer and each member of the recording scheme can obtain the following documents: a list of undisposed ewes and rams; an undisposed lamb list; a stock list; a ewe performance list; a ewe index;

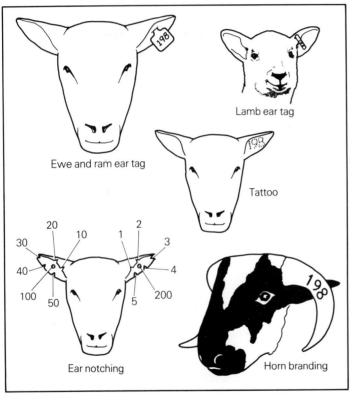

Figure 13.8 Methods of sheep identification

lamb performance list; ram performance summary; pedigree record; ewe register.

Selection based on performance recording

The service provided by the MLC Sheep Improvement Services includes the calculation of a 'Ewe litter weight index' based upon the lifetime records of individual ewes. This is presented in three parts: the first is a list of ewes in ages; the second is a list of ram lambs with the litter weight index of their dams; the third is the same but for ewe-lambs. Ewes, ram-lambs and ewe-lambs which have the highest performance indices can therefore be selected for breeding.

Other performance criteria may be used as the basis for selection, and objective information recorded in this way provides the basis for a breeding programme for commercial traits.

Co-operative breeding (group breeding schemes)

Co-operation between breeders with similar objectives enables an effective selection programme to be carried out with a larger population than is possible with individual flocks. Group breeding schemes have been successfully developed in New Zealand where at least 26 such schemes were operating in

1974. They were started by farmers trying to produce 'easy-care' sheep with good commercial performance. The schemes are a practical method of breeding, run under commercial conditions by farmers who decided to pool their resources and experience.

The basis of a group breeding scheme is simple recording of a number of flocks (ideally at least 10) which identifies individuals of superior merit. Superior ewes from the co-operating flocks are selected annually for transfer to a central 'nucleus' flock (see Fig. 13.9).

The nucleus flock comprises ewes drawn from all co-operating flocks kept under standard management and subjected to more intensive recording. The nucleus flock is maintained by selecting 50 per cent of the gimmers each year from all ewes in the nucleus flock and a regular intake of a 5 per cent selection from all ewes in co-operating flocks. Ram-lambs are selected from those produced by ewes with the highest lifetime records. About 7 per cent of the rams from ewes with the highest ratings are retained for use in the nucleus flock and a further 50 per cent of rams from the higher rated ewes are returned to the co-operators. A *pro rata* exchange of rams for draft ewes is usually practised, but the group may also have the capacity to sell rams to other breeders.

It is usually desirable to aim for uniformity of type and appearance in selected stock and this has been done in New Zealand by holding open days at the nucleus farm when final

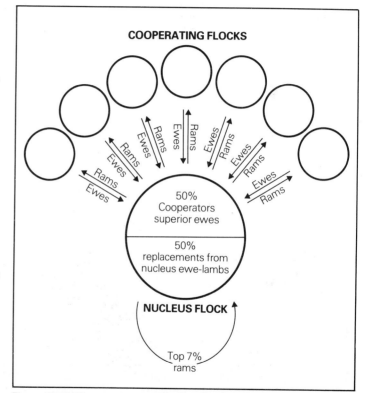

Figure 13.9 The organisation of a group breeding scheme

selection of rams (from those selected on performance records) is undertaken by farmers.

Advantages of group breeding schemes

Group breeding allows intense selection (because of the large numbers of animals) for traits of commercial and economic value under commercial conditions. More detailed and accurate selection can be carried out in the nucleus flock and special tests (e.g. progeny testing) can be carried out on the nucleus farm. All the replacement rams for the co-operating farms come from the nucleus flock so the improvements are quickly spread throughout the group and inbreeding problems are avoided.

Other benefits of belonging to a co-operative scheme are that group discussion and pooling of expertise can lead to general improvements in managemant. The group can also afford to employ technical help and advice on both management and veterinary aspects. The group also enjoys a larger corporate identity which can benefit the marketing of both slaughter lambs and breeding stock.

Schemes have been successfully started in the UK with the Welsh Mountain, Llŷn and Cambridge breeds. Undoubtedly the most likely interest will come from commercial flock-owners who find similar producers with common interests and who wish to produce rams selected under their own environment, which cannot be found from normal sources and ram breeders. One possible application is for a group of commercial farmers to co-operate with a ram breeder who would be well suited to running the nucleus flock. Otherwise the co-operators must jointly rent a farm or one co-operator must agree to take the flock onto part of his farm.

Health control

A potential problem of group breeding schemes is the transfer of infectious diseases. Screening of co-operating flocks by an independent veterinary surgeon should be undertaken at the outset and only flocks with a sufficiently high health status included. Thereafter an agreed programme of preventive medicine should be implemented by all the flock-owners. Early diagnosis of disease problems in any flock is essential and stock transferred between farms should be isolated for at least one week to observe their health before mixing with the recipient flock.

Selection

As in all breeding schemes, selection should be for a minimum number of characters of economic importance, characters with high heritability and characters which can readily be measured. Environmental differences should be minimized while ensuring that the direction of change will not be detrimental in the commercial environment either because of genotype/environment interaction or genetic correlation between the chosen factor and another important factor.

Figure 13.10 Open field day at New Zealand Romney Group Development Ltd., the first group breeding scheme in New Zealand (by courtesy of Tony Parker)

Planning the sheep system

14

A successful sheep enterprise depends on both physical and economic planning. The appropriate system must be chosen for the farm, the best breed or cross used and management and nutrition organized to ensure the maximum possible performance from the system. On lowland farms, sheep are usually subsidiary to arable cropping and, on most upland farms, cattle are also kept and a small acreage of cereals may be grown. The sheep enterprise cannot be looked at in isolation from other stock and crops, but must be integrated to allow efficient use of resources (land, labour and capital) and to maximize the total farm income. On hill farms, sheep are frequently the major enterprise and the performance and profitability of the sheep mainly determine the farm income.

In order to plan the sheep system, records of important physical and financial parameters are needed. The performance of the flock can then be compared with targets set according to the type of sheep and the management system, The records must be critically assessed to find areas for potential improvement if output and profit are to be maximized. Before making changes in the system or increasing or decreasing the size of the flock, an estimate must be made of the probable physical and economic consequences both to the sheep and to the farm as a whole.

Physical performance

Accurate information on all the inputs and outputs of the flock allow an assessment to be made of the strengths and weaknesses of management and nutrition. Records should be kept of the flock numbers and ages, the number of replacements kept or bought in, the number of draft ewes sold and the number of ewes culled or died. The most important records are those of the numbers of ewes mated and the corresponding numbers of lambs born, weaned and sold. These permit the calculation of the percentage of barren ewes, the lambing percentage, weaning percentage and selling percentage which are fundamental measures of flock performance. A measure of lamb performance can be obtained by recording the dates and weights of lambs sold, without the need to gather the flock in order to weigh all the lambs. It may be valuable to weigh lambs at the time when the first ones are expected to be marketable and to draw lambs at the same time as checking overall performance. The total weight of wool sold is provided in the annual return from the British Wool Marketing Board or its equivalent.

Condition scoring of ewes was recommended as the best method of assessing the adequacy of feeding and a note of the condition of ewes and particularly the number of lean ewes (below score $2\frac{1}{2}$) is useful. All feed inputs should be recorded, preferably on a weekly basis, and the dates and actions taken for regular preventative medicine should be included.

The sale dates and numbers of breeding stock sold should also be recorded and a regular record of all deaths of both breeding stock and lambs kept together with details of the cause of death.

Prices and costs

Along with the physical records, the sale prices of all lambs and breeding stock sold should be noted and the buying prices of replacement stock and costs of feeds and other variable costs accounted. The price of wool will be shown on the wool cheque and Government subsidies may also be payable. Many of the figures will be kept in the farm accounts, but these should be extracted for the purpose of planning the sheep enterprise and expressed on a per head basis to allow comparisons to be made.

A typical set of record sheets containing all the likely relevant information on the physical and financial performance of a commercial sheep flock are shown in Tables 14.1, 14.2, 14.3 and 14.4.

Table 14.1 Commercial flock record 1

Replacement purchases

Date	Number	Class (ewe-lambs, 18-month ewes, rams)	Price per head (£)	Haulage	Source

Autumn health measures

Date	Treatment	Product used	No. of sheep treated	Dose	Total	Cost

Numbers and condition scores of ewes before mating

Date	Age	Breed	Number	Average condition score	Number of lean ewes

Mating dates

	Ewe-lambs	Ewes
Date rams in		
Date rams out		

Rams

Identification	Breed	Age	Price when bought	Source	Fertility test Date	Result

Winter survival of ewes (mating to lambing)

	Ewe-lambs	Ewes
Number mated		
Number barren		
Number died		
Number culled		
Number lambed		

Table 14.2 Commercial flock record 2

Number and condition scores of ewes 6 weeks before lambing					
Date	Age	Breed	Number	Average condition score	Number of lean ewes

Winter feed				
Week	Number of ewes	Concentrates fed (kg)	Hay fed (kg) (bales)	Other feeds (kg)

Prelambing health measures						
Date	Treatment	Product used	No. of sheep treated	Dose qty	Total	Cost

Total consumption:	Tonnes	Price per tonne
Concentrates		
Hay		
Other feeds		

Lambing record*

	Ewe-lambs	Ewes
Date started		
Date finished		
Number of lambs born		
No. survived to 48 hours		
Number weaned		
Number sold		

*A separate lambing book is kept to record the number of ewes lambed, number of lambs born and number surviving each day.

Table 14.3 Commercial flock record 3

Spring and summer health measures

Date	Treatment	Product used	No. of sheep treated	Dose	Total	Cost

Concentrates fed to lambs

Week	Concentrates fed (kg)	Total consumption (tonnes)	Price per tonne

Lamb weights

Date	Field or group	Average weight (kg)	Weight gain since birth (av. wt. 5 kg)	Age (days)	Liveweight gain (g/day)

Lamb sales

Date	Number	Store or fat	Weight Live	Weight Dead	Price Per kg	Price Per head	Sub-sidy	Haulage/commission

Wool sales

Class (lambs, ewes)	Number clipped	Weight of wool	Grade	Price per kg	Total income

Subsidies

Number of ewes eligible	Level of subsidy	Total income

Table 14.4 Commercial flock record 4

Autumn health measures

Date	Treatment	Product used	No. of sheep treated	Dose	Total qty	Cost

Draft/cast ewe and ram sales

Date	Number	Correct in teeth and udder or fat	Price per head	Haulage/commission

Ewe and ram deaths

Date	Identification	Age	Diagnosis

Lamb deaths

Date	Identification	Age	Diagnosis

Forage costs and stocking rate

The costs of growing grass for sheep grazing and hay production include those of the seeds and fertilizer for initial establishment (which are divided by the number of years of production to give an annual cost), annual fertilizer and sprays. The grazing cost per ewe is the forage cost per hectare divided by the number of ewes grazed per hectare (stocking rate). The total forage costs per ewe must include the costs of growing the hay for winter feed and a proportion of the costs of any other pasture used by the ewe flock or fattening lambs at other times of year besides the summer grazing season.

Unfortunately, it is not always possible to allocate the forage costs so simply because the sheep enterprise and other livestock may share the same pasture, for example with mixed grazing. In these cases the forage costs can be apportioned to the different enterprises on the basis of livestock units. Comparative livestock units for different classes of stock are listed in Table 14.5. If there are 200 ewes and 60 autumn calving suckler cows on 50 ha of grass, the total livestock units are 40 + 60 = 100 LU. The costs of 20 ha should be charged to the sheep enterprise as this represents 40 per cent of the grazing livestock.

The costs of permanent pasture and rough grazing are much less than those of sown pasture but the grazing value is less. In calculating stocking rates, the area of lower quality pasture is adjusted by a factor proportional to its grazing value. A factor of 0.5 may be used for poor permanent pasture and 0.25 for rough grazing when calculating the total area of grass used by the flock. These factors are best assessed for individual farms; good permanent pasture with adequate fertilizer use can be just as productive as many sown pastures.

The grazing and hay costs are additional variable costs to the sheep flock. The summer stocking rate is the number of ewes per hectare on the summer grazing area. The stocking rate per forage hectare (annual stocking rate) is the number of ewes on each hectare used by the flock during the year, including grazing, hay and the production of other winter feeds.

Table 14.5 Comparative grazing pressures (livestock units)

Cattle		
Dairy cows	1.0	
Beef cows (excl. calf)	0.8	⎤
Calf under 6 months	0.2	⎦ 1.0
Other cattle under 300 kg	0.4	
Other cattle over 300 kg	0.6	
Sheep		
Lowland ewes + lambs	0.2	
Hill ewes + lambs	0.12	
Ewes replacements 18-month		
(lowland)	0.14	
(hill)	0.1	
Store lambs, short keep	0.02	
Store lambs, long keep	0.04	
Other sheep	0.1	

Calculating profitability

The accepted method of expressing the profitability of an enterprise for comparative purposes is the gross margin (GM). This is the output less the variable costs and may be expressed per ewe or per forage hectare. Gross margins for all the enterprises may be summed to give the *total farm gross margin*, but this is not the profit made by the farm. *Farm profit or loss* is the total farm GM less the *fixed costs*, which are the rent, labour costs and the depreciation on capital equipment. Fixed costs may also include bank interest, paid management costs, ownership costs and unpaid (family) labour. Fixed costs vary considerably between farms and are not always related to the size or performance of the enterprises. When comparing a sheep enterprise with targets based on other flocks' performance, it is misleading to include fixed costs which may be totally different on other farms. Fixed costs of a given farm may be apportioned to the different enterprises when comparing their profitability, but a decision to change the balance or abandon an enterprise on this basis can be unrealistic. It would be wiser to consider the fixed costs which would be saved by abandoning the enterprise. For these reasons, fixed costs should not be included when comparing the performance of sheep enterprises, although they be a considerable factor in the decision to start a flock.

Gross margin is therefore the most useful figure for comparing flock performance with standard targets or other flocks. A *net margin* figure may also be calculated as the GM less the *interest on working capital* (the value of the stock + half of the variable costs). This takes into account the cost of the capital directly employed in the sheep. If 100 ewes are valued at £5000 and the total variable costs are £1400, the working capital is taken as £5700. The interest at 15 per cent per annum would be £855 (£8.55 per ewe). Profitability can also be expressed as the *percentage return* on *working capital* (net margin/working capital × 100%). Figures which include the cost and return on capital are appropriate for comparison of two enterprises on a farm; in the case of cattle and sheep, cattle may produce a higher GM per hectare than sheep but

Table 14.6 Calculation of target and actual gross margins for a lowland sheep flock

	Target (per 100 ewes)	Actual (per 100 ewes)
Physical data		
Lambs born	200	......
Lamb weaned	180	......
Draft ewes sold	20	......
Replacements bought (18-month ewes)	25	......
Number of rams	3	......
Wool sold (kg)	300	......
Concentrates fed (tonnes)	5	......
Summer grazing (ha)	6	......
Hay (ha)	1	......
Swedes (ha)	1	......
Other feeds/rented grazing	0	......

Table 14.6   Calculation of target and actual gross margins for a lowland sheep flock (cont.)

	Target (per 100 ewes)	Actual (per 100 ewes)
Output	(£)	(£)
180 lambs at £29 per head	5 220	. . . . . .
20 draft ewes at £40 per head	800	. . . . . .
300 kg wool at 110p/kg	330	. . . . . .
Less: 25 replacement ewes at £70	−1 750	−. . . . . .
1 replacement ram at £200	− 200	−. . . . . .
Total output	4 400	. . . . . .
Variable costs		
Concentrates fed at £90 per tonne	450	. . . . . .
Veterinary and medical costs	150	. . . . . .
Miscellaneous costs	100	. . . . . .
Forage costs: grazing at £80/ha	480	. . . . . .
hay at £100/ha	100	. . . . . .
swedes at £120/ha	120	. . . . . .
Total variable costs	1 400	. . . . . .
Gross margin per 100 ewes	3 000	. . . . . .
Interest on working capital (£6 200 at 15%)	− 930	−. . . . . .
Net margin per 100 ewes	2 070	. . . . . .
Percentage return on working capital	33%	. . . . . .

involve a much higher capital investment and produce a lower percentage return on capital. Thus, where capital is limiting, it may be better to start or increase a sheep enterprise than to invest in cattle.

The calculation of a GM for a lowland sheep enterprise is shown in Table 14.6, using the relevant data from the commercial flock records. The output term includes the lamb, breeding stock and wool sales less the actual replacement costs; the change in valuation of the stock is sometimes included in the replacement costs, but in times of inflation this gives a much lower figure and can be misleading as it does not represent an actual cash return to the farm.

Comparison of physical and financial performance with targets

To assess the performance of a sheep flock, the physical records and GMs may be compared with performance targets for that system. In the UK, the performance of average and best producers is published each year for different systems of sheep production by the MLC and local targets are provided by agricultural advisers from ADAS and the Scottish Agricultural Colleges. Alternatively, targets may be set for the individual farm by the farmer or his adviser. Table 14.6 also shows how the performance of an example farm is compared to the target. This indicates where improvements may be made in the current performance and profitability.

Examples of target lamb weaning percentages are given for four sheep-production systems in Table 14.7. These targets have been achieved by actual flocks in the East of Scotland. They are obtained by a combination of breeding, correct nutrition and good management, including effective preventive medicine.

Table 14.7 Targets for four sheep-production systems

System	Lambs born/ 100 ewes lambed	Lambs weaned/ 100 ewes mated	Stocking rate (ewes/ha forage)
Early lamb production with Dorset–Finn ewes synchronized to lamb in August	180	160	20
Intensive lamb production with crossbred ewes on lowland pasture	200	180	13
Intensive lamb production with crossbred ewes on good upland pasture	180	160	10
Extensive lamb production with purebred ewes on hill pasture (with improved land and two-pasture system)	130	110	–

Improving the performance of a flock

When the performance of a sheep flock is below the target, the records must be analysed to find where the weaknesses are in the system and action taken to improve results.

A high proportion of barren ewes may indicate fertility problems in the rams. Physical examination and semen testing (particularly of young and recently purchased rams) may be undertaken by a vet if a problem is suspected. One ram is required for every 40–50 ewes and more (1 : 10) if ewes are synchronised. Infertile ewes should be culled to avoid repetition of the problem in subsequent years.

If the lambing percentage is low, a better result may be obtained by changing to another more productive breed or cross (see Table 14.8). The condition of the ewes at mating affects ovulation rate and better body condition at this time could improve lamb numbers born. If lamb mortality is excessive the answer lies in better nutrition of the ewes before lambing (especially the lean ewes), lambing management and effective disease prevention and hygiene. When later lamb deaths deplete the numbers sold, attention to disease problems are required. The involvement of the veterinary surgeon in diagnosing the causes of death and planning the preventive medicine programme can be invaluable (see Table 14.9). This applies equally to ewe losses and infertility problems which may be caused by inadequate nutrition, disease or ram sterility. The vet may arrange to carry out a semen test on

Table 14.8 (a) Average performance of commercial crossbred ewes in lowland flocks (b) Performance of purebred ewes in upland flocks (MLC (1978) *Sheep Facts*)

(a)

	Weight (kg)	Live lambs born/ewe lambed	No. lambs reared
Border Leicester × Blackface	70	1.79	1.50
Welsh Halfbred	58	1.56	1.36
Scottish Halfbred	77	1.76	1.49
Suffolk × S. Halfbred	80	1.70	1.37

(b)

	Weight (kg)	Live lambs born/ewe lambed	No. lambs reared
Scottish Blackface	52	1.40	1.20
Welsh Mountain	38	1.23	1.06
Clun Forest	61	1.52	1.31

rams, especially those recently purchased or ram-lambs, to establish whether they are fertile.

If lamb growth is below target it may be explained by poor nutrition of the ewes or lambs (inadequate grass availability), trace-element deficiency or (most likely) a parasite problem, often indicated by unthrifty lambs and scouring (diarrhoea).

The implementation of a clean grazing system with timely use of anthelmintics is the best solution.

Lower lamb sales returns per ewe may be due to lower numbers or a lower price per lamb. The latter could be improved by better marketing. The choice of time of sale, weight and condition are all important, but considerable variation can exist between market outlets such as meat wholesalers, live auction marts, fat and store sales. Prices for draft ewes are also subject to seasonal variation and the recent trends may indicate a more profitable alternative to present practice.

Replacement costs are partly a reflection of the flock wastage and partly the length of the productive life of the ewe within the flock. Because there is often a good market for younger ewes for breeding, it does not always pay to keep ewes to a greater age. The price of old cull ewes sold for slaughter and the price of younger ewes which are correct in teeth and udder must be compared with the cost of buying the appropriate number of replacements. Ewes may be purchased at 6 months of age (ewe-lambs) or 18 months (gimmers); 18-month ewes currently cost about 50 per cent more. The ewe-lambs either incur a cost of maintenance for one year or can be mated to produce lambs. Although it is argued that lambing ewe-lambs increases the lifetime performance of the ewes, they produce less lambs in their first year compared to their second year of life. There is a negative difference in output between ewe-lambs and gimmers, and over the same number of

Table 14.9 A routine health programme for a spring lambing flock

Before mating	Vaccinate replacement ewes for enzootic abortion* and clostridial diseases (7 in 1). Check ram fertility – especially young replacements Drench all ewes for worms and liver fluke.*
After mating	Winter dip all sheep to control lice, keds and ticks* and against sheep scab. Vaccinate replacement ewe lambs for louping ill.*
Mid-pregnancy	Give copper injection or drench ewes to prevent swayback.* Vaccinate against orf.*
Late pregnancy	Supplement feeding with minerals/vitamins. Booster clostridial vaccination to ewes 2 weeks prelambing. Observe ewes for lambing sickness and pregnancy toxaemia and consult vet to obtain appropriate remedies.
Lambing	Ensure good hygiene and watch for scour in young lambs – move to clean field or treat with antibiotics. Use antibiotic injection or pessary following difficult lambing. Dress navels with antiseptic (iodine). Supplement colostrum when necessary by stomach tube.
After lambing	Feed magnesium-rich minerals to avoid staggers. Dose ewes for worms and move to clean pasture if possible.
Early summer	Dose lambs for *Nematodirus* if not on clean pasture.* Repeat dosing for worms every 3–4 weeks in this case.* Dose hill lambs at marking and clipping. Vaccinate lambs for clostridial diseases (7 in 1) at 8–10 weeks of age.
Late summer	Summer dip for maggot fly and avoid grazing wooded areas. Drench for liver fluke.* Dose lambs for worms at weaning; move to clean aftermath.
Autumn	Dose finishing lambs for worms and fluke* and vaccinate for clostridial diseases. Give cobalt bullets or treat pasture with cobalt sulphate to prevent cobalt pine.*

*Treatment for these diseases is only required in affected areas. The disease should be diagnosed by the vet before treatment is given.

The incidence of specific diseases varies between years and between localities. The local vet should be invited to assist in planning the routine health programme for the individual farm. An early diagnosis should be sought when unexplained symptoms and losses occur.

reproductive cycles the ewe-lamb produces less (see Table 14.10). This difference must be balanced against the difference in price at purchase. The incidence of difficult births and mismothering is also far greater in ewe-lambs and this imposes a management burden at lambing time which may be unacceptable in large flocks with limited labour. Ewe-lambs are also slower to mate, require separate feeding in winter and separate grazing in summer so that the lambs can be creep fed. Without these precautions, the subsequent performance as a ewe at 2 years old may be adversely affected. There is a stronger case for lambing ewe-lambs in self-replacing breeding flocks in favourable environments. In commercial flocks on farms where there is an area of grazing unsuitable for ewes and lambs, replacements may be purchased more cheaply as ewe-lambs which are not mated and maintained at low cost for one year. This is not worth while where they compete with productive ewes for the use of good pasture.

Table 14.10 Comparison of the performance over 4 years of ewe-lamb and 18-month ewe replacements mated in the first year (after W. Rutter (1975) *Sheep From Grass* East of Scotland College of Agriculture, Bulletin 13)

	Ewe lambs			18 month ewes		
	Lambs weaned	Indiv. carcase weight (kg)	Total carcase weight (kg)	Lambs weaned	Indiv. carcase weight (kg)	Total carcase weight (kg)
First year	1.0	16	16	1.6	18	29
Second year	1.6	18	29	1.8	20	36
Third year	1.8	20	36	1.8	20	36
Fourth year	1.8	20	36	1.8	20	36
Total	6.2		117	7.0		137

At present (1980) prices the extra 20 kg of lamb carcase is worth £30; the difference in price between ewe-lambs and 18-month-old ewes is also about £30; they should be worth approximately the same if sold as draft ewes in the fourth year; extra management input is required for feeding and lambing the ewe-lambs.

Feed costs are related to ewe and lamb performance; a saving in feed cannot be tolerated when this will reduce output. However, when ewes are in good condition before lambing the level of concentrate feeding can be reduced and this may also be achieved by making better quality hay. Separating the leaner ewes and feeding more to these while feeding slightly less to the majority of fit ewes is a more efficient use of a given input of feed.

Savings on veterinary and medicine costs, which are a small proportion of total costs, are likely to represent a false economy if disease problems are increased.

Forage costs per ewe are related mainly to fertilizer inputs and stocking rate. Higher stocking rates require a greater use of fertilizer per hectare but this does not necessarily mean a greater forage cost per ewe. Grass production is directly related to nitrogen application and the amount required is proportional to the number of ewes per hectare. Thus, fertilizer costs are the same on a per ewe basis at higher stocking rates; the same amount of nitrogen is applied to a smaller area and this can release land for other uses (Fig. 14.1).

Increasing stocking rate

Animal performance may be high in relation to the target, but the GM per hectare is less because the stocking rate is low. An increase in stocking rate must not result in an appreciable reduction in animal performance or no net advantage will accrue. On lowland and upland farms, poorer performance at higher stocking rates is often attributable to an increased worm parasite burden; the introduction of a clean grazing system permits an increase in stocking rate without a reduction in ewe and lamb performance and the targets shown in Table 14.7 (p. 178) have been achieved by flocks using the system. With clean grazing and a proportional rise in nitrogen

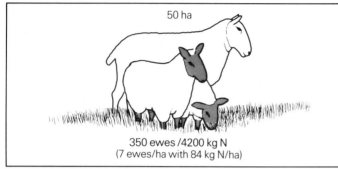

350 ewes /4200 kg N
(7 ewes/ha with 84 kg N/ha)

Low stocking rate – all grass used

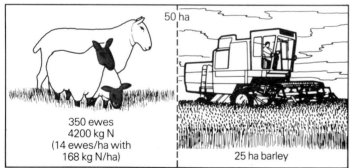

350 ewes
4200 kg N
(14 ewes/ha with
168 kg N/ha)

25 ha barley

High stocking rate – land released for barley

Figure 14.1 Comparison of fertilizer usage and output of ewes at two stocking rates

fertilizer usage, many farms could increase the summer stocking rate of sheep and output per hectare.

An increase in summer stocking rate does not result in a proportionate increase in annual stocking rate because although the grazing area per ewe is reduced, the same amount of land is required per ewe for the production of winter feed. The effect of increasing summer stocking rate on annual stocking rates is shown in Table 14.11. Higher stocking rates also involve a greater capital investment per hectare. Realistic target stocking rates for sheep can be related to potential crop yields (see Table 3.1, p. 30). These levels have been shown to work well in practice and are considerably above current UK practice. On productive lowland pasture, a stocking rate of 18 ewes/ha with the use of about 220 kg N/ha is the current maximum in use in the East of Scotland. Good ewe and lamb performance has been achieved regularly at this level, but it remains to be seen if higher stocking rates can be achieved with more fertilizer usage up to the levels employed in intensive dairy farming. The greater use of clover in pasture may reduce the requirements for nitrogen, especially at lower stocking rates, but the advantage over grass is less at high levels and the practicality of high output from grass/clover swards has yet to be demonstrated in the UK although it is the basis of many sheep systems in New Zealand. The greater seasonality of clover production results in less production in the spring in the UK.

Table 14.11 Relationship between summer stocking rate and annual stocking rate allowing for winter feed production

Summer stocking rate (ewes/ha)	5	10	15	20
Hectares required/ 100 ewes for grazing	20	10	6.6	5
Hectares required/100 ewes for winter feed production (hay/swedes)	2	2	2	2
Annual stocking rate ewes/forage ha	4.5	8.3	11.6	14.3

Whole-farm implications

Increasing the stocking rate of sheep can have several effects on the farm: a greater number of sheep can be kept or the same number can be kept on a smaller area, releasing land for other uses. With clean grazing and higher fertilizer use, more ewes can be grazed per hectare and greater output can be obtained either through more sheep, more cattle or more crop production.

The decision on which direction to increase depends partly on the relative profitability of the enterprises. In recent years, cereal production has been the best option and a marginal increase involves very little capital investment. The fertility of the land limits the extent of the increase in crop production and the inclusion of a grass break in the arable rotation is necessary on all but the best land. When cereals are most profitable, the area should be maximized to the limits of the suitable land on the farm. On upland farms, the choice is more likely to be between sheep and cattle. Again the decision can be based on the relative profitability, but a balance is necessary to allow clean grazing for both sheep and cattle. Furthermore, it is wrong to base the decision on short-term economic changes because changes in scale are not achieved overnight. A balanced mixed farm is less likely to suffer from occasional variations in the economics of an individual enterprise.

Land, labour and capital constraints also operate on any one enterprise. The maximum cereal acreage may be limited by the land or on large farms by the capacity of the machinery and labour. Similarly, the size of the cattle enterprise is often limited by the facilities for wintering (especially where winter housing is required). The sheep enterprise is eventually constrained by the capacity of the shepherd. An increase in one enterprise can be justified up to the limit where further investment and fixed costs are incurred due to the requirement for another man, additional buildings or equipment.

Table 14.12 shows the effects of intensifying the sheep enterprise on whole-farm output on seven farms in the eastern Borders of Scotland between 1974 and 1977. The farms changed from a low stocking rate with mixed grazing to a high stocking rate on the clean grazing system and as a result were able to increase ewe numbers, cattle numbers and cereal area,

Table 14.12 The effect of sheep intensification on total farm output on seven farms in S.E. Scotland (1974-77)

Sheep stocking rate	1974	1977	Per cent increase
	7 ewes/ forage ha	10 ewes/ forage ha (clean grass)	
Average farm size (ha)	221.5	221.5	–
Area of cereals (ha)	41.3	51.0	23.5
Number of ewes	420	508	21.0
Number of cows	87	100	14.5

giving an overall increase in output of about 20 per cent with no appreciable increase in fixed costs. On individual farms, the increase in particular enterprises varied with the steeper upland farms concentrating on sheep and cattle and lowland farms growing more cereals with the same number of livestock.

Improving output and profitability of hill farms

Sheep are the main enterprise on the majority of hill farms and their performance determines the profitability of the whole farm. Sheep performance is restricted by nutrition and particularly by the limit of production of hill pasture; output could be greatly increased by improving the land and its vegetation, controlling grazing to allow better utilization of improved pasture and more supplementary feeding in winter (see Ch. 4). Hill land is very variable in nature and the strategy for a particularly farm depends on the soil type, existing vegetation and current sheep management and performance.

The problem on many hill farms is that the current level of output does not generate enough profit to provide large amounts of capital for major land improvement schemes. Investment must be accompanied by a continuous gradual increase in output to service the capital required for improvement and the project must be carefully planned to match cash inputs with output to ensure an acceptable profit margin every year.

The programme of improvement described in Chapter 4, starting with an extensively managed, open-hill farm, is first to improve winter nutrition, then to achieve better control at lambing by fencing a lambing paddock, then to establish areas of improved pasture (which may be achieved on better hill land by fencing and the application of lime and slag in the first instance) and finally to carry out more land improvement (and possibly the use of more sophisticated techniques of reseeding). This programme requires the investment of capital over a period of years which is accompanied by progressive increases in output; improvement is achieved first by more lambs being produced per ewe and then, as improved pasture becomes available, by higher lamb weaning weights and the capacity to increase the number of ewes.

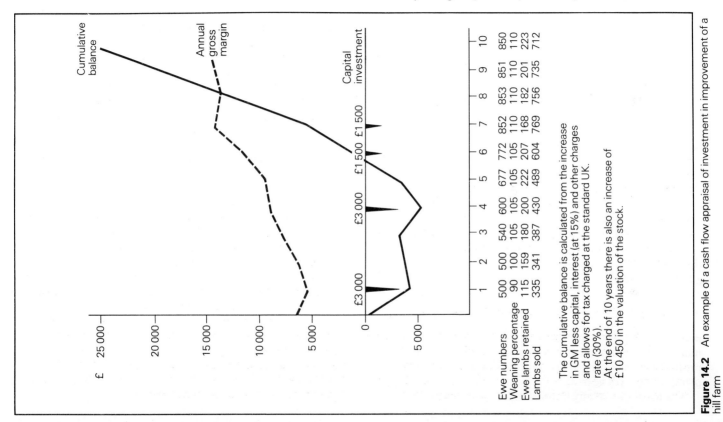

	1	2	3	4	5	6	7	8	9	10
Ewe numbers	500	500	540	600	677	772	852	853	851	850
Weaning percentage	90	100	105	105	105	105	110	110	110	110
Ewe lambs retained	115	159	180	200	222	207	168	182	201	223
Lambs sold	335	341	387	430	489	604	769	756	735	712

The cumulative balance is calculated from the increase in GM less capital, interest (at 15%) and other charges and allows for tax charged at the standard UK rate (30%).
At the end of 10 years there is also an increase of £10 450 in the valuation of the stock.

Figure 14.2 An example of a cash flow appraisal of investment in improvement of a hill farm

Increasing ewe numbers in a self-replacing hill flock can be achieved either by keeping more ewe-lambs or keeping back some of the draft ewes. With a limited number of ewe-lambs available as replacements, keeping a higher proportion means that the standard of selection is lowered and this could be detrimental to long-term performance; keeping back old ewes could result in greater management problems and mortality; both options reduce the stock sales from the farm in the short term. The increase in stock numbers must be gradual so that no more than 30 per cent of the ewe-lambs remaining – after normal replacements are selected – are kept.

When planning an improvement scheme for a hill farm, the cash flow (the movement of funds through the business over a defined period and represented by the balance of cash inputs and outputs) needs to be calculated to find whether the interest on capital borrowed to finance the scheme can be paid by the increased output and the capital repaid in a reasonable time to satisfy the lender. An example of a typical cash flow in a hill farm improvement scheme is shown in Fig. 14.2 In the example, two areas of 15 ha each are improved in years 1 and 4 with the investment of £3 000 in each year and further improvements are made in years 6 and 7 with a further investment of £3 000 over 2 years. By gradually increasing ewe numbers and the number of lambs sold from the farm, profitability is improved and the investment starts to show a profit in the sixth year.

The feasibility of any improvement scheme must be assessed in this way for the individual farm. In every case, the extra lamb output produced by increasing ewe numbers and individual performance must pay for the land improvement and other costs as the scheme progresses and eventually show a profit.

At the end of the scheme, the output of the farm and value of the stock will be considerably higher than at the start; the potential for increasing sheep production from the hills in this way is very great.

Outlook

15

Sheep production can be considerably increased without radical changes in current production methods. The potential lambing rate, growth performance, stocking rate and feed utilization are far above that presently being achieved. If more flock-masters succeed in emulating the management and methods of top producers, a substantial increase in national meat production would be achieved. The main improvements will come from better planning, attention to detail and good stockmanship.

The most important sheep-farming areas will continue to be the hill and marginal lands where there are limited agricultural alternatives, and the most profitable systems of production will be those which make efficient use of pasture and forage crops. Increasing production from pasture requires a better understanding of soils, plants and parasites, as well as of animals themselves. In the development of more intensive use of upland and hill land, the economic implications to the individual farmer, sources of finance and return on investment must also be considered. In practice, this requires close co-operation among scientists, advisers, farmers and shepherds.

The increasing shortage and price of petrochemicals brings into question the continuing use of large quantities of nitrogen fertilizer (which is a by-product of the oil industry). The greater exploitation of the nitrogen fixation of clover in pasture remains a real possibility in sheep grazing systems. This has been the focus of development in New Zealand but there remain considerable problems in growing and utilizing clover in northern Europe and other temperate regions where the growing season is short and the contribution of clover less dramatic. Again the problem must be approached from the systems viewpoint to fit the best animal system, grazing techniques and management programme to the production pattern of clover-based pasture. A similar approach could be used for developing systems using dry-land forage plants such as lucerne (alfalfa) in arid areas.

The sheep is an inefficient user of cereal grains and expensive protein, compared to pigs, poultry and dairy cows, and as such would be a poor competitor with non-ruminants for limited resources of feed grains and certainly with the use of cereals for direct consumption by humans. It is unlikely that the availability of cereals and proteins for animal feed will increase or that the cost will decrease in the future and, more probably, greater pressure on crop production to feed the increasing world population can be expected.

Much attention has been paid in recent years by scientists and development agencies to the possibility of intensifying sheep production by increasing output per ewe. In countries as far apart as the US, Israel, Finland, Spain, Iran, France and Ireland, systems of production involving three lambings in 2 years and highly prolific ewes capable of producing over two lambs in a single crop have been attempted. Success has depended on the skill of expert technicians and the input of much greater amounts of expensive feeds. Intensive methods such as these usually make poor use of natural resources.

There are situations where intensive production methods can be complementary to extensive methods, for example, in feedlot finishing of lambs from breeding flocks maintained under more extensive conditions. Integration of intensive and extensive production systems in this way could improve the utilization of national resources, particularly in arid areas. Intensive systems may also play a small role in developed countries where technical expertise exists, the price of lamb is high and the local availability of cereal grains makes their use economic for meat production, but this is untrue of vast areas of the world where labour is less skilled and crop production is already insufficient to provide for existing local demands.

There is still much to learn if we are fully to exploit the potential of existing sheep breeds for greater lamb production, growth performance and lean meat yield. Looking to the future, we may well see changes in the requirements for carcase size and lean meat content. These changes are already occurring in the UK with a demand for small lean carcases by the European market. However, greater efficiency could be achieved by producing *large* lean carcases and the use of such products could be evolved by modifying present techniques of meat cutting, presentation and marketing. Newer methods of meat handling and promotion are already being tried in the United States, and adoption of these marketing ideas could influence future approach to market requirements and production systems for sheep meat.

Effective marketing can also greatly improve the

profitability of sheep products for the farmer. An increase in farmer-controlled, co-operative selling of the products could have a very beneficial effect both on the financial return and on the feedback of market information to the production level. Co-operation would also allow more efficient and less costly transportation and handling of products.

Changes in labour availability and costs are likely to have increasing effects on sheep production. Greater simplicity and ease of management are prerequisites of sheep-production techniques under conditions of reduced labour input. There is also a need to train more shepherds and sheep farmers to cope with the demands of increased flock sizes, modern production methods and changing market requirements. Effective training is essential for the successful adoption of new ideas and increased output from sheep flocks.

The greater involvement of veterinarians in sheep farming would also help to alleviate the considerable waste of animals from disease. But vets must have a thorough understanding of production systems and financial constraints if they are to offer practical solutions to the commercial farmer.

It will be clear that a wide field of knowledge is required by those involved in developing and applying new methods in sheep production. We need greater exchange of ideas between those developing and applying new production techniques and more attention to the whole system and whole farm implications of new ideas.

Further reading

H.F.R.O. *Hill Farming Research Organisation Reports*, Bush Estate, Penicuik, Midlothian, EH26 0PY, Scotland.

McDonald P., Edwards R.A. and Greenhalgh J.F.D. (1973) *Animal Nutrition*, (2nd edn), Longman, London and New York.

M.L.C. (1978) *Sheep Facts: a manual of economic standards*, Meat and Livestock Commission, P.O. Box 44, Queensway House, Bletchley, Milton Keynes, MK2 2EF, UK.

Owen J.B. (1976) *Sheep Production*, Bailliere Tindall, London.

Tomes G.L., Robertson D.E. and Lightfoot R.J. (eds) (1979) *Sheep Breeding* (2nd edn, revised by W. Haresign), Butterworths, London.

Watt J.A. (1971) *The Shepherd's Guide: a guide to the diseases of sheep and to the care and training of sheepdogs*. Advisory Bulletin no. 9, Dept. Agric. Fish., Scotland. H.M.S.O., London.

Index

Ind